# Genetics and Morphogenesis
# in the Basidiomycetes

# Genetics and Morphogenesis in the Basidiomycetes

*Edited by*

## MARVIN N. SCHWALB

*Department of Microbiology*
*CMDNJ-New Jersey Medical School*
*Newark, New Jersey*

## PHILIP G. MILES

*Department of Biology*
*The State University of New York at Buffalo*
*Buffalo, New York*

ACADEMIC PRESS   New York   San Francisco   London   1978

*A Subsidiary of Harcourt Brace Jovanovich, Publishers*

ACADEMIC PRESS, INC.
111 Fifth Avenue, New York, New York 10003

*United Kingdom Edition published by*
ACADEMIC PRESS, INC. (LONDON) LTD.
24/28 Oval Road, London NW1 7DX

LIBRARY OF CONGRESS CATALOG CARD NUMBER:

ISBN 0-12-632050-0

PRINTED IN THE UNITED STATES OF AMERICA

*To the memory of John Robert Raper,*
*scientist, teacher, friend.*
*Not so much the words or the paper, but the ideas.*

# Contents

# Contributors

Numbers in parentheses indicate the pages on which the authors' contributions begin.

PETER R. DAY (67), The Connecticut Agricultural Experiment Station, 123 Huntington Street, New Haven, Connecticut 06504

Y. KOLTIN (31), Department of Microbiology, Faculty of Life Sciences, Tel-Aviv University, Ramat Aviv, Israel

PHILIP G. MILES (1), Department of Biology, The State University of New York at Buffalo, Buffalo, New York

DONALD J. NIEDERPRUEM (105), Department of Microbiology, Indiana University School of Medicine, Indianapolis, Indiana 46202

CARLENE A. RAPER (3), Department of Biology, Harvard University, Cambridge, Massachusetts

MARVIN N. SCHWALB (135), Department of Microbiology, CMDNJ-New Jersey Medical School, Newark, New Jersey 07103

JUDITH STAMBERG (55), Microbiology Department, Faculty of Life Sciences, Tel-Aviv University, Tel-Aviv, Israel

J. G. H. WESSELS (81), Department of Developmental Plant Biology, Biological Centre, University of Groningen, Haren, Nederland

# Preface

When we first conceived of this symposium we wanted the presentations to consist of more than a recitation of the most recent results. Rather we asked each participant to summarize and analyze the research covered by their topics. Furthermore, we encouraged the expression of new ideas. We believe that our goals have been met and that this volume represents more of the future than the past.

We thank our many colleagues who provided their thought and time. The officials of the Second International Mycological Congress should be congratulated for a well-run meeting. Our special thanks to Eloise Henry and Paula Shatten for their efforts on the IBM Recorder and Composer.

# Genetics and Morphogenesis
# in the Basidiomycetes

# INTRODUCTION*

Philip G. Miles

Although officials of the Second International Mycological Congress could not give official approval to a memorial symposium, the fact is that this symposium on Genetics and Morphogenetic Studies of Basidiomycetes had its inception in a desire to honor John Robert Raper. Scores of scientists were contacted, and there was overwhelming approval of the idea of a symposium of this congress in memory of Professor Raper. Suggestions were also sought as to the topics to be covered in the symposium, but on this there was more divergence of opinion, as would be expected in view of John Raper's many scientific contributions and broad interests which included the hormonal control of sexual development in fungi, the biological effects of beta radiation, and the genetic control of the incompatibility systems and morphogenesis of sexuality in higher basidiomycetes. We incorporated as many suggestions as possible, but ultimately Dr. Schwalb and I had to make the decisions as to the contents and organization of this symposium, recognizing that it could have been done in many different ways. We are confident, however, that any imperfections in the organizational pattern will be less obvious because of the contributions of this outstanding group of participants.

Mrs. Raper expressed to me some concern that she had been invited to be a participant simply because she was John Raper's wife. Dr. Schwalb and I assured her that she is on the program because of her own specific scientific contributions in this field and that it would not in our opinion have been in keeping with John's strict sense of scientific honesty to have selected a participant for reasons other than that the person was an outstanding representative for the topic to be covered. As John's wife alone, we would have honored her with a front seat at this symposium, but not a place on the program.

Many of you are familiar with John Raper's scientific

* This statement was read at the beginning of the Symposium

accomplishments and some will be familiar with various phases of his career. For those who are not familiar with this, I would like to give a brief synopsis.

The youngest of eight children, John Raper was born on a farm in North Carolina. He received both the bachelor's and master's degrees from the University of North Carolina, an institution well known to mycologists for the studies of Coker and Couch and their students. What more auspicious introduction to mycological science could one have than to learn about fungi from Professor John Couch? The next stop was Harvard University for another master's degree and the Ph.D. under the tutelage of Professor W.H. Weston, Jr., a recognized master in the training and development of young biological scientists. This was followed by a two year post-doctoral fellowship at Cal Tech where he grew *Achlya* in great quantities and isolated a small but significantly useful amount of hormone A. His first teaching position was at Indiana University, and this was interrupted by a period of research on the Manhattan Project at Oak Ridge where he studied the biological effects of beta radiation. Following the war he took up a position at the University of Chicago where he continued the *Achlya* studies and embarked on the investigations of the genetic control of incompatibility in the tetrapolar basidiomycete, *Schizophyllum commune,* while climbing up the academic ladder from Assistant to Full Professor. In 1954 he became Professor of Botany at Harvard University to succeed "Cap" Weston. While he did not match in numbers Cap's record of guiding over 50 students to the Ph.D., the number of outstanding young scientists whose research was guided by John marks an outstanding contribution he has made to our profession.

The spirit of his active inquiring mind will be very much with us during these meetings, for many of you here have known him and few could meet him even casually without sensing that he was an extraordinary man. If he were here with us physically today, I think that he would be getting a bit restless by now and would probably be saying: "There are excellent people waiting to tell us some exciting things. Let's get on with it." So be it!

# CONTROL OF DEVELOPMENT BY THE INCOMPATIBILITY SYSTEM IN BASIDIOMYCETES

Carlene A. Raper*

## INTRODUCTION

Many aspects of development in Basidiomycetes fascinated John Raper, but the question of central importance to him was, always, "How do the incompatibility genes do their work?"

Despite the decades of effort resulting in considerable information about the incompatibility system -- a signifigant part of which was contributed by John Raper -- the answer to this question is little more apparent now than it was over fifty years ago when, independently, Marie Bensaude (1918) and Hans Kniep (1920) first defined the system in the higher Basidiomycetes. The answer is of relevance, not only to an understanding of mating interactions and sexual development throughout the higher fungi, but also to an understanding of the control of development in eukaryotes generally. Specific analogies in genetic aspects are apparent in the $S$ allele system for control of fertilization in higher plants and in the control of histocompatibility in higher animals.

In essence, the products of the incompatibility gene-complexes interact to convert the fungus from one state of differentiation to another through a sequence of morphologically distinct events. The system has been detected in approximately 90% of the Homobasidiomycetes analyzed. This represents about 450 heterothallic and secondary homothallic species of the estimated 5000 species extant (Raper, 1966).

Although research to date has not revealed the nature of the products of the incompatibility genes, the accumulated information makes speculation about this question more tempting than ever. I, as

* Department of Biology, Harvard University, Cambridge, Massachusetts, U.S.A.
The recent studies of C.A. Raper and J.G.H. Wessels reported here were supported by the Netherlands Organization for the Advancement of Pure Research (ZWO). In memory of my husband, John R. Raper.

others previously, succumb to that temptation, with a consideration first of relevant background information, then an explication of two proposed models for the molecular basis of incompatibility gene function, followed by a discussion of these models as they accommodate known facts and as they might be tested.


## BACKGROUND

The detailed sequence of events in development as controlled by the genes of the incompatibility system varies among species of Basidiomycetes but a common feature is the conversion of a self-sterile homokaryotic mycelium to a heterokaryon that is capable of forming fruiting bodies.

A generalized scheme of the life cycle of Basidiomycetes is given in Figure 1. Its salient features are the alternation of an indefinite haploid phase with a one-celled diploid phase, usually with a heterokaryotic phase of dikaryotic structure interposed. Under appropriate environmental conditions, the heterokaryon is induced to produce fruiting bodies containing basidial cells in which karyogamy, meiosis and spore formation occur in rapid succession.

The illustrated scheme typifies events in the majority of known species, such as *Coprinus fimetarius* (Bensaude, 1918; Mounce, 1922), *Schizophyllum commune* (Kniep, 1920), *Lentinus edodes* (Oikawa, 1935; Nisikado and Yamauti, 1935), *Flammulina velutipes* (Kniep, 1920; Vandendries, 1923), and *Pleurotus ostreatus* Vandendries, 1933). (See Raper, in press, for comprehensive references on these species). Variations in several of the steps are found in other species. For example, in *Agaricus bitorquis,* the homokaryotic mycelium (step 2) is multikaryotic instead of monokaryotic and the fertile heterokaryotic mycelium (step 4) is dikaryotic without clamp connections (Raper, 1976). In *Agaricus bisporus,* a secondary homothallic form, the fertile heterokaryon not only has no clamp connections but is multikaryotic. There is no apparent dikaryotic structure except for cells just basal to the basidia. Also the spores from the two-spored basidia are dikaryotic at conception with the consequent bypassing of the homokaryotic

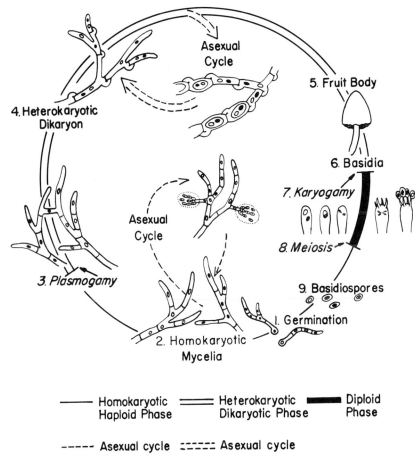

Fig. 1    A generalized scheme of the life cycle of Basidiomycetes.

phase (step 2) (Raper et al, 1972). *Armillaria mellea* represents still another variant in its regular sectoring of the dikaryon into monokaryotic mycelium with presumably diploid cells, and without clamp connections (Korhonen and Hintikka, 1973, J.B. Anderson and R.C. Ullrich, personal communication). A small percent of the species studied have no alternation of the self-sterile, homokaryotic phase with the heterokaryotic phase but are true, so-called primary, homothallics, in which the homokaryon is self-fertile and an incompatibility system is lacking. Morphology of the fruiting body varies widely from the unelaborate, as in resupinates such as *Sistotrema brinkmannii,* to the elaborate, as in Agarics such as *Amanita sp.,* and asexual cycles of various types may or may not occur in the homokaryotic and heterokaryotic phases.

These variations in the expression of development, however, do not obscure the signifigance of the genes of the incompatibility system as the primary controlling elements, and it is this I wish to focus on.

Sexual morphogensis in Homobasidiomycetes is controlled by extensive series of multiple alleles of either a single incompatibility factor, $A$ or of two incompatibility factors, $A$ and $B$. Because the details of its operation have been examined most comprehensively in the bifactorial species, *Schizophyllum commune* Fr., I will review the relevant information available from studies with this organism as a background for discussion.

The $A$ and $B$ incompatibility factors are unlinked and complex genetic factors that control distinct but coordinated parts of the sexual cycle. In bifactorial species such as *Schizophyllum,* the $A$ and $B$ factors segregate at meiosis to produce four types of basidiospores with respect to incompatibility genotype: *Ax Bx, Ay By, Ax By, Ay Bx*; Each type develops into a self-sterile, cross-fertile homokaryon which, when paired in matings with the other three types gives three distinct patterns of sexual morphogenesis. The entire progression leading to the development of the fertile dikaryon (*A-on B-on*) occurs only when there are allelic differences in both $A$ and $B$ factors (e.g. *Ax Bx* X *Ay By*). Only part of the series of events (*A-on*) occurs when the $A$'s are different but the $B$'s are the same (e.g. *Ax Bx* X *Ay Bx*): and only another part (*B-on*) occurs when the $B$'s are different but the $A$'s are the same (e.g. *Ax Bx* X *Ax By*). The *B*-sequence of events (*B-on*) involves the reciprocal exchange and migration of nuclei into and throughout the mycelium of each mate and, at a later stage, the fusion of hook cells (clamps). The A-sequence of events (*A-on*) involves the pairing of nuclei, one from the donor mycelium and one from the acceptor mycelium, in each cell, the formation of the specialized hook cell at each septum, conjugate nuclear division, and the septation of hook-cell and hypha immediately following nuclear division. (See Raper, 1966 for further details.) Fruiting normally occurs only when both the A- and B-sequences are operating (*A-on B-on*) (see Schwalb, this symposium). The events and their genetic controls are illustrated in Figure 2.

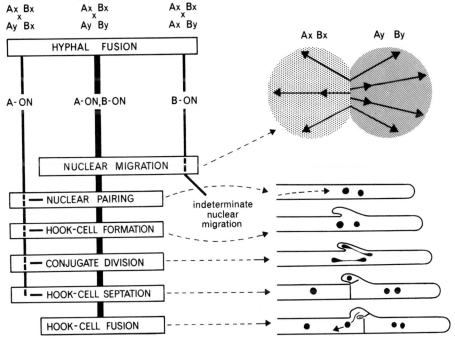

Fig. 2. Control by the $A$ and $B$ incompatibility factors of sexual morphogenesis in *Schizophyllum*. The progression comprises two distinct and complementary series of events, the A-sequence and the B-sequence, traced by vertical lines at left and right and regulated by the $A$ and $B$ factors, respectively. Operation of either sequence requires interaction of compatible factors, e.g. *Ax Bx* X *Ay Bx* which is *A-on*, or *Ax Bx* X *Ax By* which is *B-on*. Morphogenesis is completed only when both sequences are operative as traced by the central, heavy, vertical line, *Ax Bx* X *Ay By* which is *A-on, B-on*. (From Raper, J.R. and C.A. Raper. 1973. Brookhaven Symp. Biol. No. 25).

Distinct physiological and biochemical characteristics have been identified as correlates of these three patterns of morphogenesis (see Wessels, this symposium). Differences in the spectra of total soluble proteins were first shown to correlate with sexual morphogenesis by serological tests (Raper & Esser, 1961) and later by polyacrylamide gel electrophoresis (Wang & Raper, 1969). Different isozyme patterns for many enzymes were also demonstrated to be associated with specific states of sexual differentiation (Wang & Raper, 1970).

In the *B-on* phenotype, metabolism is shifted to catabolism with the concomitant elevation of several hydrolytic enzymes (Wessels & Niederpruem, 1967; Wessels, 1969), and energy conservation is highly inefficient (Hoffman & Raper, 1971, '72, '74). The production of large amounts of a specific hydrolytic enzyme known as R-glucanase has been shown to be essential for the dissolution of septa, a necessary prelude to the continuous movement of nuclei from cell to cell that is characteristic of the *B-on* phenotype (Wessels & Marchant, 1974). A preliminary study indicates the involvement of cyclic AMP in the formation of the specifically differentiated hook cells in the *A-on B-on* phenotype (Gladstone, 1973).

The primary genetic controls of this pleomorphic array of characteristics are the four loci of the two independently assorting incompatibility factors, *A* and *B*. Each factor is constituted of two linked genes, $\alpha$ and $\beta$, and each gene has multiple states with respect to specificity. In the *A* factor there are 9 and 32 alleles of the $A\alpha$ and $A\beta$ genes respectively (Raper et al, 1960; Stamberg & Koltin, 1973) and in the *B* factor there are 9 alleles each of $B\alpha$ and $B\beta$ (Parag & Koltin, 1971). There is no evidence that the products of the two genes within a factor interact with each other but they do appear as functional equivalents. A mating between two haploid homokaryons heterozygous at either or both of the loci within a factor results in the *on* phenotype for that factor – a minimum of a single difference at either locus versus no difference at both loci within a factor has widely different developmental consequences. It is the combined specificities of both $\alpha$ and $\beta$ genes that confers specificity to the factor. The *A* factor therefore has an estimated 288 specificities and the *B* factor 81 making a total of approximately 23,328 mating types with respect to combined $A$ and $B$ genotypes.

Extensive multiple specificities for the incompatibility factors in both unifactorial and bifactorial forms is common throughout the Homobasidiomycetes. For example, the unifactorial form of *Sistotrema brinkmannii* has an estimated 100-300 $A$ specificities (Ullrich & Raper, 1974) and the bifactorial *Pleurotus ostreatus* has an estimated 63 $A$ specificities and 190 $B$ specificities (Eugenio & Anderson, 1968). A complex structure for the single incompatibility factor in unifactorial forms has not yet been demonstrated, but recent studies with *Agaricus bitorquis* revealed not only multiple allelism for its single incompatibility factor but indications of recombination for mating type specificity within the factor (Raper, 1976).

The genetic components of the incompatibility system have been studied primarily through mutational analysis. A study of mutations affecting the expression of morphogenesis has revealed, in addition to the incompatibility genes, a large array of loci scattered throughout the genome that determine specific aspects of the developmental process. They are expressed only during sexual morphogenesis and are viewed as secondary controls of the process. They are recognized in impairment as "modifier" mutations and constitute that component of the system that is regulated by the incompatibility genes. Over 80 mutations representing 12 phenotypes with respect to specific effects on morphogenesis have been analyzed. Several are expressed as specific blocks or alterations to the A-sequence of morphogenesis; others to the B-sequence and some to both sequences (Raper & Raper, 1964, '66, and unpublished). Most appear to be not linked to the incompatibility genes, but a cluster of nine modifier mutations expressed as blocks to nuclear migration in the *B-on* phenotype are linked by 10-20% recombination with the $B$ factor (Dubovoy, 1975).

The complex nature of the regulatory components has become apparent from the variety of mutations obtained in the incompatibility loci. Although exhaustive attempts to derive one allele from another through mutation have failed, many phenotypes with respect to alterations in the control of sexual morphogenesis have been generated by mutations within the locus. Mutations have been obtained in three loci, $A\beta$ $B\alpha$ and $B\beta$, with the latter most intensively studied. In a combined sample of approximately $10^9$, some 50 mutations in both $B$ loci were obtained as a primary event. They occur with the application of various mutagenic agents in a frequency of $5 \times 10^{-8}$ and are of two types regardless of locus. Both are constitutive for the operation of the B-sequence of morphogenesis ($B$-on) but one has retained its ability to recognize the parental type as identical to itself and the other has lost that ability (Raudaskosky et al, 1976, and Koltin, this symposium). Mutations as secondary events in the $B\beta$ locus, i.e. mutations of a primary $B\beta$ mutant, are 1000 times more freqeunt than primary mutations and represent at least 10 types. All express a $B$-off phenotype and have varying degrees of deficiency in effecting the $B$-on phenotype when combined with wild type alleles in matings. Among these deficiencies are failures in the acceptance of nuclei, the donation of nuclei, and hook cell fusion. The mutant phenotypes range all the way from reversion to the parental wild-type allele to complete lack of all functions even extending to the adjacent $B\alpha$ locus. All of these secondary mutants, including the latter, are recessive to the parental, constitutive, primary mutant (Raper & Raper, 1973).

The variety of discernable alterations achieved in this single incompatibilty locus indicates a complex gene of two major parts, one for specificity concerned with self versus nonself recognition in allelic interaction, and one for the function of initiating and regulating the B-sequence of sexual morphogenesis. The primary mutations are interpreted as alterations within the specificity region and the majority of the secondary mutations are interpreted as alterations within the regulatory region, sometimes extending into the specificity region. Both regions appear to be subdivided. The specificity region has at least two parts, one for allelic interactions, and one for permitting expression of the regulatory region once a nonself allelic interaction has occurred. In the primary mutations obtained, the latter is always altered to permit constitutive expression; the former is sometimes retained as the parental type and sometimes destroyed. The region for regulatory function appears to be subdivided into three parts, one for the acceptance of nuclei, one for the donation of nuclei, and one for the fusion of hook cells. The function for acceptance of nuclei in a mating takes precedence over the other two functions.

Serious attempts to resolve some parts of an incompatibility locus through intragenic recombination have succeeded only in separating that part of the $B\beta$ locus responsible for regulating nuclear acceptance. The mutated form, termed $f$ for functionless, is separable from the remainder of the locus in a frequency of 0.8%. No recombination for allelic specificity has been detected in any of the incompatibility loci despite attempts in pooled samples of $10^7$ by several investigators (J.R. Raper & Baxter, C.A. Raper, Koltin, Raudaskosky, all unpublished).

Another aspect of possible relevance to an understanding of the function of the incompatibility loci is their apparent association with deletions of varying sizes. An analysis of recombinational spectra within the $B$ factor has demonstrated the inability of some $\alpha$ and $\beta$ alleles in the natural population to combine with one another. Overlapping of deletions is invoked as the probable basis for this phenomenon (Stamberg & Koltin, 1971). This hypothesis is supported by results of a comparable analysis of recombination between the natural $B\alpha$ alleles and some secondary $B\beta$ mutations. No recombination could be achieved between those $B\alpha$ alleles thought to have large deletions and the more severely altered $B\beta$ mutations. A logical interpretation is that these particular secondary mutations are deletions overlapping those deletions associated with the natural and completely functional $B\alpha$ alleles. In fact, the secondary mutant with no function for either $\alpha$ or $\beta$ cannot be recombined with any $\alpha$ allele. It appears as a deletion spanning both $\alpha$ and $\beta$ loci (Stamberg & Koltin, 1974; Koltin, this symposium).

The initiation of morphogenesis by the incompatibility genes, whether by products of a constitutive mutation in a homokaryon, or by a nonself recognition interaction between products of unlike alleles in a wild type heterokaryon, requires a considerable lag period. A period of 58 hours, at the optimal temperature of 30°C, is required for initiation of the B-sequence (septal dissolution and the movement of nuclei between cells) in developing germlings of basidiospores carrying a constitutive $B$ mutation (Koltin & Flexer,

1969). Similarly, the initiation of the A-sequence (nuclear pairing and hook cell formation) occurs after a lag period of 60 hours in germlings with constitutive mutations in both the *A* and *B* factors (Koltin, 1970).

In a fully compatible mating between mycelia of wild-type homokaryons, the first signs of nuclear migration appear from 24-48 hours after mating (Snider & Raper, 1958; Raudaskosky, 1973; Leary & Ellingboe, 1970), and the entire process of nuclear migration requires 60-72 hours before any evidence of the A-sequence is apparent. A recent preliminary study (C.A. Raper & Wessels, unpublished) comparing morphogenesis in regenerating protoplasts of various origins has revealed an even longer lag period of about 90 hours for the initiation of nuclear pairing and hook-cell formation in developing regenerates of fusion cells between protoplasts of fully compatible homokaryons. By contrast, regenerating protoplasts from the dikaryon and from homokaryons with constitutive mutations in the *A* factor and/or *B* factor express either all or part of the relevant morphogenesis almost immediately, from the first few regenerating hyphal cells onward. Regenerating protoplasts from a mutant constitutive for both the A and B-sequences (*A-on B-on*) start hook-cell formation between the first and tenth cell generations; at first the hook cells are predominantly unfused, but the proportion of unfused to fused hook cells is about the same after the tenth cell generation as it is in the aged mycelium. Regenerating protoplasts from a mutant constitutive for the B-sequence (*B-on*) express part of the *B-on* phenotype (irregular branching, occasional disrupted septa and irregular nuclear distribution) beginning at about the tenth cell generation, and half of the regenerates from a mutant constitutive for the A-sequence form hook cells (nucleated pseudoclamps) consistently within the first few cell generations.

Regenerating protoplasts of the dikaryon are predominantly heterokaryotic, and these, at first, form either fused hook cells (true clamps) at each septum or unfused hook cells (pseudoclamps); by the 15th cell generation, true clamps and the dikaryotic phenotype prevail and persist. A smaller proportion of protoplasts from the dikaryon, 10-40%, are homokaryotic and sometimes, but not always, represent both of the two nuclear types. About half of the homokaryotic regenerates differentiate hook cells transiently, even though dissociated from the unlike incompatibility genes that initiated this form of differentiation. In other words, there is a lag period for reversion to the stable morphology of the wild type homokaryon once the "genetic trigger", at least at the nuclear level, is removed. The expression is variable: pseudoclamps develop in average from the 10th to 30th cell generations and then are no longer formed (Wessels et al, 1976).

This phenomenon, first observed by Harder (1927) in homokaryons microsurgically derived from dikaryons of *Schizophyllum commune* and later confirmed by Lange (1966) in similar experiments with dikaryons of *Polystictus versicolor,* raises questions as to the basis for the persistence of cell differentiation in the absence of its initiating genetic control. Harder suggested a cytoplasmic determinant of long lasting activity. Wessels proposed a nuclear basis, in which both nuclei in dikaryotic association acquire the new differentiative function to initiate hook-cell formation and this function persists for varying lengths of time after dissociation. He argues that the delay in expression of hook cells in homokaryotic regenerates may be explained by assuming destruction, through experimental manipulation, of a cytoplasmic condition needed for expression of this nuclear activity; after repair through growth, the necessary cytoplasmic condition may be restored and the nuclear function is expressed until its activity dissipates in the absence of the "genetic trigger", i.e. unlike incompatibility genes. The latter hypothesis has some support from the results of Lange (1966) in which five of the nine homokaryons isolated from the dikaryon expressed hook-cell formation and were of one nuclear type, while the other four, which did not express hook-cell formation were of the other nuclear type. This suggests that the differentiative function is determined primarily not in the cytoplasm but is correlated with genomic differences in the two component nuclei.

A very recent attempt to clarify this has "muddied the waters" even further. The results from analysis of large samples of homokaryotic regenerates from protoplasts of a dikaryon of *Schizophyllum* remained inconclusive and at the same time revealed an unexpected phenomenon (C.A. Raper & Wessels, unpublished). In a sample of 264 homokaryons dissociated from a particular dikaryon, 24% developed no hook-cells during regeneration and the remaining 76% expressed a range in hook-cell formation from a very low to very high incidence. As in the earlier studies, hook-cell formation was transient, occurring from about the 10th to 30th cell generations. The incidence of hook-cell formation could not be correlated with nuclear type. In fact, the suprising result of this study was the highly disparate ratio of nuclear types that emerged: 262 isolates were of one nuclear type from the dikaryotic pair and only 2 were of the other.

The basis for this phenomenon, verified in four experiments with three independently established dikaryons from the same two compatible strains, is not known. Results so far suggest that it cannot be explained on the basis of one genome (of the minority nucleus) having acquired a recessive debilitating mutation: protoplasts from the two component, unmated strains regenerate in equal frequency under identical conditions, and progeny (single spore isolates) from the dikaryon survive in high frequency, appear normal, and reveal the expected ratio for segregation of incompatibility types. Nor does it appear to be a "maternal" effect: subcultures from both sides of the mating produce the same results. It is clearly a case of a discriminatory effect in the dikaryon upon the ability of one member of the dikaryotic pair to survive independently of the other immediately after separation.

In this particular dikaryon, one member assumes dominance over the other. Such clear dominance must not prevail in all dikaryotic associations in view of the fact that Harder (1927), in his microsurgical experiments, recovered both nuclear types in small samples from two of four different dikaryons. It is interesting to note, however, that Harder was unable to recover one nuclear type from similar experiments with *Pholiota mutabilis*.

Further work will be directed towards establishing a possible genetic basis for this phenomenon and a determination of its incidence in other dikaryons of *Schizophyllum*. Whether or not it relates to the operation of the incompatibility system is not yet known, but it shows that nuclei can acquire some difference by being associated in a dikaryon.

## MODELS FOR MOLECULAR BASIS

Any speculation about the molecular basis for the operation of the incompatibility loci must take into account those salient facts at hand. They are, in summary:

1. Each incompatibility locus has an extensive number of alleles.
2. For a given locus, the gene product of each allele is capable of distinguishing the product of an identical allele (self) from those of the many other nonidentical alleles (nonself), and this recognition phenomenon is mediated through the cytoplasm.
3. For a given locus, the consequence of nonself recognition is pleomorphic and apparently identical regardless of the specificity of the alleles involved, e.g., the consequence of interaction between products of $B\beta 1$ and $B\beta 2$ is identical to that of interaction between products of $B\beta 3$ and $B\beta 4$.
4. Initiation of the morphogenetic process by nonself recognition requires a signifigant lag period, and also the reversion of at least one morphogenetic process after removal of the genetic initiators requires a lag period.
5. All alleles are strictly equivalent in their differences from one another; there is no overlapping complementation in which the product of one allele recognizes those of two others as self but those two recognize each other as different.
6. New alleles cannot be derived by intra-allelic recombination in a frequency of greater than $10^{-6}$.
7. One allele cannot be derived from another by induced mutation in a frequency of greater than $10^{-9}$.
8. Alleles are associated with blocks, of various degrees, to recombination. The blocks appear to be deletions.
9. The incompatibility locus is functionally complex as indicated by, a) primary mutations, constitutive for regulation of morphogenesis, that either have or have not lost the parental specificity for recognition, (induced in frequency of $5 \times 10^{-8}$), and b) secondary mutations of malfunction expressing various

quantitative and qualitative impairments in recognition of other allelic specificities and in regulation of morphogenesis, (induced in frequency of $7 \times 10^{-5}$). The constitutive mutants are dominant to the malfunctioning mutants, and a constitutive mutant can be reversed. A part of the locus, regulating one of at least three discernable regulatory functions, is separable by recombination.

10. The expression of many loci throughout the genome is regulated by the incompatibility genes, and mutations in these loci, modifying the normal course of morphogenesis in a variety of ways, are induced in a frequency of $2 \times 10^{-4}$.

The products of the incompatibility genes must, of necessity, be complex and discriminatory in order to accomodate these known facts about the loci and their functions. They must have two major functions, one for the discrimination between self and many other similar products and one for the initiation of the revelant sequence of morphogenesis. Proteins alone may perform both functions, as suggested by Kuhn and Parag (1972), or a nucleic acid may be involved in the discriminatory function, as suggested by Ullrich (1973).

To simplify discussion of the models put forward by these authors, we will consider the two loci of a given factor as having identical but independent functions -- there is no evidence to the contrary -- and we will consider the alleles at a single locus of one well studied incompatibility factor only, e.g. the $B\beta$ locus.

The Kuhn-Parag model proposes that self versus nonself recognition reflects a difference in the ability of protein molecules to interact in the cytoplasm to form aggregates which are active in the regulation of morphogenesis. A single allele in the homokaryon codes for protein subunits of a single kind. A specifically different but closely related subunit is coded for by each of the nine $B\beta$ alleles. Identical protein subunits do not form active aggregates but nonidentical subunits do. This model, in its simplest form, is illustrated in Figure 3. Its authors suggest the possibility of other, more complex protein aggregation patterns, but the model is based essentially on the priniciple of change in the state of aggregation.

They assume a limited number of recognition sites in the

PROTEIN  PRODUCTS  OF  Bʙ  ALLELES

**1**  and  **2**

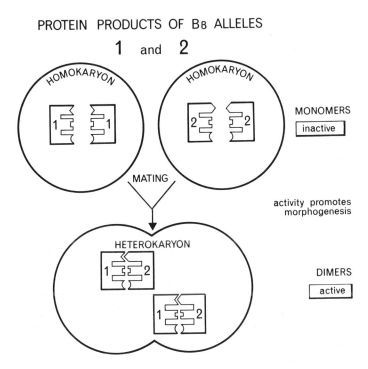

Fig. 3.    Kuhn—Parag model for protein subunit aggregation as the basis for self versus nonself recognition and the regulation of sexual morphogenesis in Basidiomycetes. The diagram illustrates the function of two alleles at one incompatibility locus only. Protein subunits coded for by the $B\beta1$ allele in the homokaryon at upper left and by the $B\beta2$ allele in the homokaryon at upper right remain unaggregated (monomeric) and consequently inactive in promoting morphogenesis. When subunits of the homokaryons are brought together in a heterokaryon through mating, aggregates (dimers) form between the two types of molecules and these are active in promoting morphogenesis. (Illustration adapted from Kuhn & Parag (1972).)

protein subunits and that a difference at a single site may be sufficient for effective aggregation. Nine alleles may represent only three sites with two alternatives or, perhaps, two sites with three alternatives. Similarly, as many as 64 alleles could be generated by six sites of two alternatives per site or by fewer than six with three alternatives.

The products of the genes are assumed to be active in the heteroallelic state. Theoretically, the reverse could be true, but the mutational evidence is against it. (An apparent deletion of the entire *B* factor is inactive with all other *B* factor specificities save those with constitutive mutations, yet the morphological appearance of such a mutant is identical to wild-type).

Kuhn and Parag interpret constitutive mutations as changes in the recognition site, perhaps in state of charge, such that self-aggregation of identical subunits to produce active molecules is no longer hindered. They interpret other mutations in the incompatibility genes, such as those represented by secondary *B* mutants that do not accept nuclei but do donate nuclei in matings, as alterations that permit self-aggregation but of an inactive type. The nonfunctional self-aggregation precludes the formation of needed amounts of active aggregate in the mutant's own cytoplasm, but the mutant subunit is capable of forming active aggregate with the wild-type subunits (which are in excess) in the cytoplasm of the compatible, wild-type mate.

The Kuhn-Parag model is broadened in concept when we consider the likelyhood that interaction of protein subunits produces conformational change and that it is the resulting change in shape of the molecule(s) that distinguishes the active from inactive form.

The Ullrich model relies on the relative complementarity of bases in a two-stranded nucleic acid to explain the recognition phenomenon. It postulates the division of the incompatibility locus into two parts, a specificity region, *S*, for self versus nonself recognition interactions, and a functional region, *F*, for synthesis of regulatory molecules responsible for the initiation of the relevant morphogenesis. The *S* region works as a molecular switch with configurations "active" or "inactive" in promoting morphogenesis. The *F* region codes for regulatory molecules that are synthesized only when the molecular switch of the adjacent $S$ region is in the active position. The $F$ region for all alleles at a given locus, e.g. $B\beta$, is assumed to be constant, whereas the $S$ region is assumed to be variable. It is further hypothesized that both strands of the entire locus are copied (transcribed or replicated) and the products of both strands are carried separately from the nucleus through the cytoplasm to a site where they are released for annealing. Products of identical alleles would be completely complementary and consequently anneal overall to form homoduplexes, whereas products of non-identical alleles would be complementary in the $F$ region but not in the $S$ region and would anneal completely only in the $F$ region to form heteroduplexes. The heteroduplexed form is the active configuration for promoting the synthesis of the regulatory molecules coded for in the $F$ region; the homoduplexed form is inactive (Figure 4). The active configuration can be thought of as exposing a promoter site for the activation of the $F$ region. The configurational difference may, of course, be more subtle than that illustrated.

The nature of the nucleic acid, whether DNA or RNA, is not specified. If it is DNA, the $F$ region of the active molecule in heteroduplexed configuration is first subject to transcription, if RNA, then it would be subject to either translation, or reverse transcription. Products of the $F$ region may be enzymes or other molecules that regulate morphogenesis at any one of a number of levels.

According to this model, the primary, constitutive mutations would represent simple nucleotide alternations that destory the secondary structure of the nucleic acid at a critical point. This could happen without changing the overall specificity of the region and would explain the constitutive mutants that retained parental specificity and, also, the reversion to parental type of a constitutive mutant that had lost its ability to recognize the parental type as identical. The secondary mutations can be explained either by simple changes in the complex functional portion of the locus, or by deletions of various sizes throughout the locus. The dominance of constitutive mutants over secondary mutants suggests a positive control by the activated $F$ region.

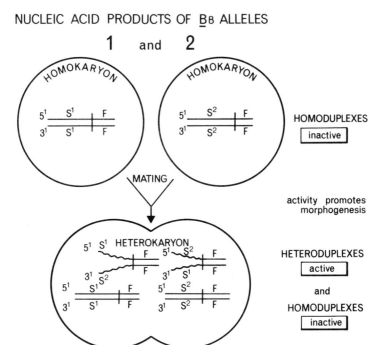

Fig. 4          Ullrich model for nucleic acid molecules as the basis of self versus nonself recognition and the regulation of sexual morphogenesis in Basidiomycetes. The diagram illustrates the function of two alleles at one incompatibility locus only. The locus is divided into a specificity region *(S)* and a functional region *(F)*, The *S* region is variable and specifies allelic differences according to nucleotide sequence. The *F* region is constant and codes for molecules initiating morphogenesis. The complementary strands of the nucleic acid product from the *B*β1 allele in the homokaryon at upper left and from the *B*β2 allele in the homokaryon at upper right are complementary in both the *S* and *F* regions and consequently form homoduplexes that are inactive in promoting morphogenesis. When the strands of the two homokaryons, which are copied and carried separately to a site of annealing, are brought together in a heterokaryon through mating, heteroduplexes form due to lack of complementarity in the *S* region and these are active in promoting morphogenesis. Homoduplexes form also in the heterokaryon but these are recessive to the heteroduplexes.

## DISCUSSION

Several known aspects of the incompatibility system can be accomodated by both models, but any comparison must be balanced against the requirements that each model specifies for the molecules involved. The Kuhn-Parag model assumes no unknown characteristics of protein activity, whereas the Ullrich model requires the undemonstrated phenomenon that each strand of the nucleic acid double helix is copied and transported separately to a site of annealing. Beyond this, the nucleic acid model has an advantage in explaining some aspects of the incompatibility system. It better accomodates the requirement, for instance, that the products of many different alleles must interact to produce a positive and identical effect, (i.e. the initiation of morphogenesis), and that there is no overlapping activity resulting from the interaction of these products. The nucleic acid model accounts for this by assuming a negative consequence of allelic interaction in the sense that the products do not complement and non complementarity promotes activity. The protein model accomodates this less easily in its assumption that allelic interaction has a positive consequence; each product complements every other product to produce an identically active molecule. Furthermore, the failure to derive new allelic specificities through mutation or intra-allelic recombination is more easily rationalized on the basis of the nucleic acid model in which allelic specificities are viewed as accumulated, multiple differences that preclude homoduplexing in a critical region of the nucleotide sequence. The generation of new alleles, at least through mutation, would be expected on the basis of the protein model in which allelic differences are thought to be inherent in one specific difference at any one of a few sites.

Other models have been proposed for incompatibility gene function: A simple complemenation of deficiencies has been considered but rejected on several grounds including the strict equivalence of natural alleles. Three models invoke repression (Prevost, 1962; Raper, 1966), and another (Pandey, 1977) is concerned with the mechanism for control of temporal events in morphogenesis. All assume proteins as the functional products and do not attempt to justify the variety of genetic facts considered here. For the earlier models, however, much of this information was not available at the time of speculation.

Still, some of the known facts are not accounted for by any of these speculations, and much remains to be learned. For example, the supposed deletions of various size associated with alleles of the $B$ factor in *Schizophyllum* obviously affect recombination between some $B\alpha$'s and $B\beta$'s, but do they have any signifigance with respect to function of the incompatibility loci? Why are so many cell generations required for the initiation of morphogenesis once the initiating genetic components (different incompatibility alleles or constitutive mutations) are introduced? And, conversely, why are so many cell generations required for the cessation of at least one morphogenetic event (hook-cell formation) when the initiating genetic components are removed?

Some questions directly relevant to tests of the two models discussed would be: 1. Can mutants temperature sensitive for the recognition phenomenon be generated and detected? If so, this would support the Kuhn-Parag model. 2. Can protein aggregates be detected by double labeling experiments in mixtures of protein extracts of two compatible homokaryons as contrasted to unmixed extracts of the differently labeled individual homokaryons? (See Kuhn & Parag, 1972, for explanation). If so, this might support the protein model, but the amounts may be so small that they cannot be detected; if detected, they may involve products of unrelated phenomena. 3. Can double stranded nucleic acids of hybrid character be detected by double labelling experiments in mixtures of nucleic acid extracts of two compatible homokaryons? If so, this would lend support to the Ullrich model, but detection and interpretation would pose similar, if not more serious, problems to those mentioned above for a test of protein aggregates according to the Kuhn-Parag model.

The most direct approach to a characterization of the product(s) of an incompatibility gene would be the sequencing of the gene. Current methods available for the isolation, recombination and amplification of a gene through cloning of bacterial plasmids, however, are not applicable to this system until an assay for the product(s) can be found. The alternative of using a closely linked marker for which there is an assay is also not now available.

The development of a bioassay for an incompatibility gene, in which the product(s) obtained in a cell free system have a measurable effect on test cells, is therefore of prime importance. So far, attempts to detect direct effects of these products in such systems have failed. The activity appears to be strictly intracellular and the molecules of interest seem to be incapable of penetrating the plasma membrane. An exception may be the effect of incompatibility factors on hyphal fusion, but the evidence for this is conflicting and difficult to interpret (Sicari & Ellingboe, 1967; Ahmad & Miles, 1970; Smythe, 1973).

A possible way of meeting this problem is through membrane fusion experiments in which the test cells are protoplasts and the cell-free fractions to be tested are packaged in membranes to form artificially made vesicles that can be fused with the test protoplasts. Such a procedure would allow the direct incorporation of active cell fractions in the absence of a membrane barrier. The effects would be detected in the regenerating recipient protoplasts.

Relevant methods for the incorporation of specific molecules in artificially generated membrane-bound vesicles and the fusion of these with viable test cells has been worked out for animal cells (Loyter, et al, 1975; Poste & Papahadjopoulos, 1976). Comparable methods should be applicable to fungal protoplasts. Methods are now available for the generation of protoplasts of *Schizophyllum*, capable of regenerating hyphal cells with normal walls, (de Vries & Wessels, 1972, 1975). Through the use of appropriate osmotic stabilizers, the protoplasts can be separated into two populations, one nucleate and one anucleate. A first step in testing the feasibility of the suggested approach would be the fusion of anucleate protoplasts of one type with nucleate protoplasts of a compatible type and the detection of any effects in regenerated products.

Protoplast fusion has already been demonstrated in fungi, (Anne & Perbedy, 1975; Ferenczy, et al, 1976). In preliminary work, we have applied a combination of methods adapted from previous work with fungal protoplasts and with plant protoplasts (Constabel, et al, 1975) to *Schizophyllum* and have achieved up to 30% fusion of nucleate protoplasts (Raper & Wessels, unpublished). A characterization of the regenerating fusion products, however, as previously mentioned, suggests that the fusion of two compatible wild type homokaryons, *Ax Bx* X *Ay By*, results in no immediate morphological expression -- in fact, hundreds of cell generations are required before any obvious evidence of morphogenetic effect becomes apparent. Possibly biochemical effects of the recognition of different incompatibility alleles occur as early events, but such are not presently detectable. The immediate expression of morphogenesis, however, in regenerating protoplasts of dikaryons or of mutants constitutive for morphogenesis, suggests the possibility of detecting effects of the regulatory molecules in such a system. The effect must first be demonstrated in cytoplasmic transplant experiments in which regenerating products of fusion between anucleate protplasts of dikaryons or of constitutive mutants and nucleate protoplasts of wild type homokaryons express transiently but immediately some morphogenetic process, such as hook-cell formation, that is inherent in the constitutive phenotype. Such results would implicate a relatively long lasting cytoplasmic element in morphogenesis. Through the appropriate use of modifier mutations and mutations in the incompatibility loci as a means of dissecting the effect genetically, it might be possible to trace the origin of effect to the product(s) of the regulatory part of the incompatibility locus. A bioassay for the effective molecular agents might then be devised by fractionating the extract from donor cells into various classes of proteins, nucleic acids, etc. and packaging the fractions in membrane bound vesicles either according to defined methods with erythrocyte ghosts or lipid vesicles in animal cell systems or by methods employing membranes derived from the fungus itself. The effectiveness of a particular fraction would then be measured by manifestation of the appropriate morphogenetic effect in the regenerating products of fusions between the loaded membrane-bound vesicles and the nucleate test protoplasts.

Perhaps this approach, even though difficult and, at the moment, highly speculative, will help bring an answer to that question that intrigued John Raper so much, "How do the incompatibility genes do their work?"

# REFERENCES

Ahmad, S.S., and Miles, P.G. (1970). Hyphal fusions in the wood-rotting funguus *Schizophyllum commune* I. the effects of incompatibility factors. Genet. Res. Camb. **15**,19-22.

Anné, J., and Perbody, J.F. (1975). Conditions for induced fusion of fungal protoplasts in polythylene glycol solution. Arch. Microbiol. **105**,201-205.

Bensaude, M. (1918). Reserches sur le cycle évolutif et la sexualité chez les Basidiomycétes. Thesis, Paris. 156 pp.

Constabel, F., Dudits, D., Gamborg, O.L., and Kao, N. (1975). Nuclear fusion in intergeneric heterokaryons, a note. Canadian Jour. Bot. **53**2092-2095.

Dubovoy, C. (1975). A class of genes affecting *B* factor-regulated development in *Schizophyllum commune*. Genetics **82**,423-428.

Eugenio, C.P., and Anderson, N.A. (1968). The genetics and cultivation of *Pleurotis ostreatus*. Mycologia **60**,627-634.

Ferenczy, L., Kevei, F., Szegedi, M., Franko, A., and Rojik, I. (1976). Factors affecting high-frequency fungal protoplast fusion. Experientia **32**,1156-1158.

Gladstone, P. (1973). Glucose repression of clamp production in diploid of *Schizophyllum commune*. Genetics **74**,595 (abstract).

Harder, R. (1927). Zur Frage nach der Rolle von Kern und Protoplasma in Zellgeschehen und bei der Ubertragung von Eigenschaften. Z. Bot. **19**,337-407.

Hoffman, R.M., and Raper, J.R. (1971). Genetic restriction of energy conservation in *Schizophyllum*. Science **171**,418-419.

Hoffman, R.M., and Raper, J.R. (1972). Lowered respiratory response to ADP of mitochondria isolated from mutant-*B* strain of *Schizophyllum commune*. J. Bacteriol. **110**,780-78.

Hoffman, R.M., and Raper, J.R. (1974). Genetic impairment of energy conservation in development of *Schizophyllum*: efficient mitochondria in energy-starved cells. J. Gen. Microbiol. **82**,67-75.

Kniep, H. (1920). Über morphologishe und physiologische Geschlechstdifferenzierung. (Untersuchungen an Basidiomyzeten.) Verh. Phys. Med. Ges. Würzburg. 46,1-18.

Koltin, Y. (1970). Development of the *Amut Bmut* strain of *Schizophyllum commune*. Arch. Mikrobiol. 74,123-128.

Koltin, Y., and Flexer, A. (1969). Alteration of nuclear distribution in *B*-mutant strains of *Schizophyllum commune*. Jour. Cell Sci. 4,739-749.

Korhonen, K., and Hintikka, V. (1974). Cytological evidence for somatic diploidization in dikaryotic cells of *Armillariella mellea*. Arch. Microbiol. 95,187-192.

Kuhn, J., and Parag, Y. (1972). Protein subunit aggregation model for self-incompatibility in higher fungi. J. Theror. Biol. 35,77-91.

Lange, I. (1966). Das Bewegungsverhalten der Kerne in fusionierten Zellen von *Polystictus versicolor*(L). Flora Abt. A. (Jena) 156,487-497.

Leary, J.V., and Ellingboe, A.H. (1970). The kinetics of initial nuclear exchange in compatible and noncompatible matings of *Schizophyllum commune*. Amer. J. Bot. 57,19-23.

Loyter, A., Zakai, N., and Kulka, R.G. (1975). "Ultramicroinjection" of macromolecules or small particles into animal cells. A new Technique based on virus-induced cell fusion. Jour. Cell Biol. 66,292-304.

Mounce, I. (1922). Homothallism and Heterothallism in the genus *Coprinus*. Trans Brit. Mycol. Soc. 7(4),256-269.

Nisikado, N., and Yamauti, K. (1935). Studies on the heterothallism of *Cortinellus Berkeleyana* Ito et. Imai, an economically important edible mushroom in Japan. Ber. Ohara Inst. Landw. Forsch. 7,115-128.

Oikawa, K. (1935). Sex in *Cortinellus shiitake*. Bot. Mag. (Tokyo) 49,453-455. (In Japanese with English summary.)

Pandey, K.K. (1977). Generation of multiple genetic specificities: origin of genetic polymorphism through gene regulation. Theor. Appl. Genet. 49,85-93.

Parag, Y., and Koltin, Y. (1971). The structure of the incompatibility factors of *Schizophyllum commune*.

Constitution of the three classes of *B* factors. Molec. Gen. Genet. **112**,43-48.

Poste, G., and Papahadjopoulos, D. (1976). Lipid vesicles as carriers for introducing materials into cultured cells: influence of vesicle lipid composition on mechanism(s) of vesicle incorporation into cells. Proc. Nat. Acad. Sci. **73**,1603-1607.

Prévost, G. (1962). Étude génétique d'un Basidiomycéte: *Coprinus radiatus* Fr & Bolt. Ann. Sci Natur.Bot. 12th Ser. **3**,425-613.

Raper, C.A. (in press). Sexuality and breeding. Chapter in "The Biology and Cultivation of Edible Mushrooms" (Chang, S.T., and Hays, W.A., eds.) Academic Press, N.Y.

Raper, C.A. (1976). Sexuality and life-cycle of the edible wild *Agaricus bitorquis*. J. Gen. Microbiol. **95**,54-66.

Raper, C.A., and Raper, J.R. (1964). Mutations affecting heterokaryosis in *Schizophyllum commune*. Amer. J. Bot. **51**,503-513.

Raper, C.A., and Raper, J.R. (1966). Mutations modifying sexual morphogenesis in *Schizophyllum*. Genetics **54**,1151-1168.

Raper, C.A., and Raper, J.R. (1973). Mutational analysis of a regulatory gene for morphogenesis in *Schizophyllum*. Proc. Nat. Acad. Sci. U.S. **70**,1427-1431.

Raper, C.A., Raper, J.R., and Miller, R.E. (1972). Genetic analysis of the life cycle of *Agaricus bisporous*. Mycologia **64**,1088-11117.

Raper, J.R. (1966). Genetics of Sexuality in Higher Fungi. Ronald Press, N.Y. 283-pp.

Raper, J.R., Baxter, M.G., and Ellingboe, A.H. (1960). The genetic structure of the incompatibility factors of *Schizophyllum commune:* the *A* factor. Proc. Nat. Acad. Sci. U.S. **46**,833-842.

Raper, J.R., and Esser, K. (1961). Antigenic differences due to the incompatibility factors in *Schizophyllum commune*. Z. Vererb. Lehr **92**,439-444.

Raudaskosky, M. (1973). Light and electron microscope study of unilateral mating between a secondary mutant and a wild-type strain of *Schizophyllum commune*. Protoplasma **76**,35-48.

Raudaskosky, M., Stamberg, J., Baunik, N., and Koltin, Y. (1976). Mutational analysis of natural alleles at the *B* incompatibility factor of *Schizophyllum commune:* α2 and β6[1,2]. Genetics 83,507-516.

Sicari, L.M., and Ellingboe, A.H. (1967). Microscopical observations of initial interactions in various matings of *Schizophyllum commune* and of *Coprinus lagopus.* Amer. J. Bot. 54,437-439.

Smythe, R. (1973). Hyphal fusions in the Basidiomycete *Coprinus lagopus* sensu Buller. 1. some of the effects of incompatibility factors. Heredity 31,107-111.

Snider, P.J., and Raper, J.R. (1958). Nuclear migration in the Basidiomycete *Schizophyllum commune.* Amer. J. Bot. 45,538-546.

Stamberg, J., and Koltin, Y. (1971). Selectively recombining *B* incompatibility factors of *Schizophyllum commune.* Molec. Gen. Genetics 113,157-165.

Stamberg, J., and Koltin, Y. (1973). The organization of the incompatibility factors in higher fungi: the effect of structure and symmetry on breeding. Heredity 30,15-26.

Stamberg, J., and Koltin, Y. (1974). Recombinational analysis of mutations at an incompatibility locus of *Schizophyllum.* Molec. Gen. Genet. 135,45-50.

Ullrich, R.C. (1973). Genetic determination of sexual diversity in the *Sistotrema brinkmannii* aggregate. Ph. D. Thesis, Harvard University, 141 pp.

Ullrich, R.C., and Raper, J.R. (1974). Number and distribution of bipolar incompatibility factors in *Sistotrema brinkmannii.* Amer. Naturalist 108,507-518.

Vandendries, R. (1923). Recherches sur le determinisme sexuel des Basidiomycètes. Acad. Roy. Belgique Classe Sci. Mém. 4[e], Sér 2, 5:1-98.

Vandendries, R. (1933). De valeur du barrage sexuel comme critérium dans l'analyse d'une sporée tetrapolaire de basidiomycète: *Pleurotus ostreatus.* Genetica 15,202-212.

Vries, O.M.H. de, and Wessels, J.G.H. (1972). Release of protoplasts from *Schizophyllum commune* by a lytic enzyme preparation from *Trichoderma viride*. J. Gen. Microbiol. 73,13-22.

Vries, O.M.H. de, and Wessels, J.G.H. (1975). Chemical analysis of cell wall regeneration and reversion of protoplasts from *Schizophyllum commune*. Arch. Microbiol. 102,209-218.

Wang, C.S., and Raper, J.R. (1969). Protein specificity and sexual morphogenesis in *Schizophyllum commune*. J. Bacteriol. 99,291-297.

Wang, C.S., and Raper, J.R. (1970). Isozyme patterns and sexual morphogenesis in *Schizophyllum commune*. Proc. Nat. Acad. Sci. U.S.A. 66,882-889.

Wessels , J.G.H. (1969). Biochemistry of sexual morphogenesis in *Schizophyllum commune:* effect of mutations affecting the incompatibility system on cell-wall metabolism. J. Bacteriol. 98,697-704.

Wessels, J.G.H., Hoeksema, H.L., and Stemerding, D. (1976). Reversion of protoplasts from dikaryotic mycelium of *Schizophyllum commune*. Protoplasma 89,317-321.

Wessels, J.G.H. and Marchant, J.R. (1974). Enzymic degradation in hyphal wall preparations from a monokaryon and a dikaryon of *Schizophyllum commune*. J. Gen. Microbiol. 83,359-368.

Wessels, J.G.H., and Niederpruem, D. (1967). Role of a cell-wall glucan-degrading enzyme in mating of *Schizophyllum commune*. J. Bacteriol. 94,1594-1602.

# GENETIC STRUCTURE OF INCOMPATIBILITY FACTORS - THE ABC OF SEX.

Y. Koltin*

The incompatibility factors of the higher basidiomycetes are unique even among the fungi. However, the solution to the problem offered by this group of organisms by the need to regulate the breeding mechanism in the absence of sexual differentiation can be seen as a classical solution to a problem in genetic organization. The same solution has been used by various organisms, both plant and animal, (de Nettancourt, 1972; Stimpfling, 1971) throughout evolution, when confronted with the necessity to evolve numerous specificities as a mean of identity in the absence of morphological differentiation.

The survival and success of any species can be perceived as a reflection on the ability of a population to regulate its short- and long-term adaptation. Sexuality as the most efficient means for the generation of genetic variation, both for genetic reassortment and the generation of additional diversity by mutations (Magni, 1964; Magni and von Borstel, 1962; Koltin, Stamberg and Ronen, 1975), can express its maximal potential only following the interaction between different genotypes. Thus, the efficiency of this mechanism is dependent on the ability to distinguish like from unlike. Sexual differentiation is clearly the most common mechanism that serves this purpose among the higher eukaryotes.

Differentiation related to the different sexes evolved in some groups of fungi but this type of morphological differentiation is totally lacking among the higher basidiomycetes. Instead, a system of recognition evolved that is based on intracellular interactions regulated by one or more genes. This system in its most complex form is probably the most flexible in the regulation of both the inbreeding and outbreeding potential of a species. The outbreeding potential is the major factor that will determine the degree of

* Department of Microbiology, Faculty of Life Sciences,
Tel - Aviv University, Ramat Aviv, Israel.

diversification of the population, whereas the degree of inbreeding determines the short range adaptation to prevailing conditions.

The higher basidiomycetes can be catagorized into 3 groups according to the organization of the incompatibility system that operates as the recognition mechanism. The 3 known groups are (a) those lacking any evident control of the mating interaction, the homothallics; (b) a group in which a single factor regulates the mating system and therefore species belonging to this group are known as the unifactorials (also known as bipolars); and (c) a group in which 2 factors regulate the mating system. The species belonging to this group are known as the bifactorials (or tetrapolars). Only one exception to this grouping has been reported: in *Psathyrella* it is claimed that 3 factors regulate the breeding system (Jurand and Kemp, 1973). This report is as yet unverified.

The relative distribution of the different groups indicates that the most prevalent are the bifactorials and the least prevalent are the homothallics. Among those behaving as homothallics only few are true homothallics and the majority behave as such due to meiotic abnormalities that lead to secondary homothallism. Based on published data for 230 species of Hymenomycetes and Gastromycetes, it is estimated that about 90% are heterothallic and some 55% are bifactorials (Whitehouse, 1949; Esser, 1967). Additional, more recent, studies of various species only tend to confirm the earlier estimates (Takemaru and Fujioka, 1969; Takemaru and Ide, 1970; Takemaru and Fujioka, 1970; Takemaru and Oh'Hara, 1970; Takemaru and Ide, 1971; Welden and Bennett, 1973; Berthet and Boidin, 1966; Boidin, 1966; Boidin and Lanquentin, 1965; Martin and Gilbertson, 1973; Roxon and Jong, 1977; Raper, 1976).

The number of genes regulating the mating interaction determine primarily the degree of inbreeding. If one gene with 2 alternate states (A1 and A2) controls the system, the progeny of any cross will consist of 2 mating types only. If mycelia from spores of one fruit body are allowed to interact freely, 50% of the interactions will be compatible. A similar situation is found in every form in which the sexes are distinct by virtue of the differentiation related to sex.

Outbreeding is influenced primarily by the number of times any individual in the population confronts a genotype identical to its own within the population. If outbreeding is of significance for the survival of the species, two evolutionary alternatives are available to the species in controlling the breeding mechanism in the absence of sexual differantiation. One alternative depends on the action of many genes, with only 3 alleles at each locus, thus forming a multitude of genotypes in which rarely rwo individuals will be alike. In this case outbreeding would be very high but inbreeding very low. An increase in the number of genes leads to a diminishing level of inbreeding. The second alternative is based on the retention of the optimal system for the regulation of inbreeding along with the evolution of a large number of states (mutations, alleles) of each gene. The system would thus compromise few genes with numerous specificities and would have the additional advantage of high outbreeding.

Among the higher basidiomycetes the second course of evolution was selected in both the unifactorials and bifactorials. The regulation of the mating system is performed by very few genes (one to four) with multiple specificities (alleles) at each locus.

In the generation of allelic specificities two opposing factors determine the extent of the allelic series, the selective advantage of each additional specificity and the limits of any gene to mutate to a new normal functional allelic specificity. In spite of the selective advantage or selection pressure posed by the relative importance of outbreeding, the generation of new functional alleles depends on the size of the locus, the nature of the gene product, etc. Thus, any locus can generate only a finite series of functional alleles. However, if the selection pressure for additional outbreeding potential is high the generation of new alleles or factor specificity may require the interplay of more than one locus to circumvent the limit set by the original incompatibility gene. The simplest type of organization beyond the one locus system to bypass the limitation posed by the individual gene is a two locus system in which the two loci jointly would detemine the factor specificity. An allelic difference at any of the loci would be sufficient to change the factor specificity. Thus, few alleles at each locus numerous factor specificities could be derived since the number of factor specificities is the multiple of the allelic specificities at each of the two loci. This type of organization introduces an additional advantage that is totally absent in systems in which each incompatibility factor consists of a single gene. When two genes determine the specificity of that factor the possibility to regulate the inbreeding potential becomes available.

Among the unifactorials the common organization is based on a one locus incompatibility factor with as many as 55 known alleles (Ullrich and Raper, 1974). The most intensive studies in any unifactorial species to detect a more complex structure were performed by Burnett in *Polyporus betulinus* (unpublished, quoted by Raper, 1966), and in *Polyporus palustris* by Flexer (1963). Both studies failed to find any indication to suggest that the incompatibility factor of these unifactorial species consists of more than one locus. The only hints that these factors may in fact be more complex stems from the study with *Sistotrema brinkmanii* (Lemke, 1969) and more recent studies in *Agaricus bitorquis* (Raper, 1976). However, attempts to repeat the results with *Sistotrema* by Ullrich (1973) have failed.

Studies on the structure of the incompatibility factors of bifactorials were initiated by Papazian (1951) in *Schizophyllum commune*. These studies were later followed by examination of the organization of each of the two incompatibility factors in *Coprinus lagopus* (Day, 1960), *Collybia velutipes, Coprinus* sp., *Lentinus edodes, Pleurotus spondelicus* and *S.commune* (Takemaru, 1961), *Pleurotus astreatus* (Terakawa, 1960) and an intensive analysis of the *B* incompatibility factor of *Schizophyllum* (Raper, Baxter and Middleton, 1958; Koltin, Raper and Simchen, 1967). Nonparental factors compatible with the parental factors were found in at least one of the 2 incompatibility factors of each of the bifactorial species studied. In bifactorial species in which nonparental factors were not detected the level of resolution (sample size) was inadequate. With the exception of *Coprinus lagopus,* all other tests were based on very small samples. Both *A* and *B* factor recombinants were found in *Pleurotus ostreatus* (Eugenio and Anderson, 1969) at a frequency of 0.5% and 4.0% after an earlier failure by Terakawa (1960). However, the positive results with the abovementioned species that included in some instances the generation of the parental factors from the nonparental factors suggest that a complex organization of the incompatibility factors in the bifactorials is typical to this taxonomically heterogenous group. Furthermore, in cases in which a complex organization was detected it appears that the factor is comprised of 2 linked loci ranging from very tight linkage of .07% in *Coprinus lagopus* (Day, 1963) to loose linkage of 23% in the *A* factor of *Schizophyllum* (Raper, Baxter and Ellingboe, 1960). The linkage relations indicate that the component loci of each factor are on the same chromosome. However, the degree of linkage was found to be variable in different crosses and was later shown to be a reflection of the fine control of genetic recombination in that specific region (Simchen, 1967; Koltin and Stamberg, 1973; Stamberg and Koltin, 1973).

As in the unifactorials each locus in the bifactorials is polymorphic. If one assumes that in most, if not all, of the unifactorials the structure of the single factor consists of one locus then each factor specificity represents an allelic specificity. The extent of the allelic series of the unifactorials equals or exceeds the number of allelic specificities found in the bifactorials. As many as 33 and 55 alleles were identified in *Fomes cajanderi* and in *Sistotrema brinkmanni* respectively (Neuhauser and Gilbertson, 1971; Ullrich and Raper, 1974). Yet, since the factor specificity in a bifactorial is determined by its component loci the few alleles of each locus jointly generate a series of factor specificities that exceeds by far any known series in the unifactorials.

In *Schizophyllum* the number of alleles at each of the 4 loci of the 2 incompatibility factors, *A* and *B*, is $A\alpha$, 32 $A\beta$, 9 $B\alpha$, and 9 $B\beta$ (Raper, Baxter and Ellingboe, 1960; Parag and Koltin, 1971; Stamberg and Koltin, 1973). The number of factor specificities is therefore theoretically 288 *A* factors and 81 *B* factors. The maximum number of identified allelic specificities in the unifactorials are ca. 55 (Ullrich and Raper, 1974) and according to the highest projected values based on few data as high as 100-300 factors. In most cases the estimated values are much lower.

Thus, the typical characteristics of the incompatibility systems of the higher fungi are: (1) regulation of breeding by few genes; in the minority of species the regulation is performed by one factor (one gene) and in a vast majority of the species the regulation is performed by 2 factors comprised of 2 genes each; (2) multiple alleles are known in each locus; (3) when the regulation is performed by 2 incompatibility factors they are unlinked but the 2 component loci of each factor are linked; (4) the specificity of a factor, in the cases in which the factor consists of two loci, is determined by the specificity of the alleles at the 2 loci.

Basically, the major difference between the unifactorials and bifactorials resides in the number of genes that regulate the system. However, the prevalence of the bifactorials suggests some selective advantage to that type of genetic organization. To assess the advantages of either type of organization the effect of the genetic organization on two aspects of the breeding system, namely inbreeding and outbreeding, must be compared.

The inbreeding potential (Table 1) in the unifactorials is fixed at 50%. If the single factor were constructed of 2 loci like the incompatibility factors of the bifactorials, the inbreeding potential would tend to rise, as a function of the linkgage relations between the two genes, up to a maximum of 75% if the 2 loci were unlinked. In the bifactorials the minimum level of inbreeding potential is 25% and the maximum is 56% if the 4 loci were unlinked. However, the outstanding fact is that in the bifactorials a fixed value cannot be determined and only a range of values can be estimated. The inbreeding potential is determined by the linkage relations of the 2 component loci of each factor and the superimposed genetic regulation of recombination in the region of each factor. In *Schizophylllum*, the highest recombination values, obtained in separate crosses, for the *A* and *B* factors allow only 44% inbreeding potential (Stamberg and Koltin, 1973). The dominance relationships of genes regulating recombination ensure that the inbreeding potential will usually be much lower, around 30%. However, the population retains the ability to increase this potential up to 44%. How critical the degree of inbreeding is to the survival of the species is difficult to assess. However, the general limit set by both unifactorials and bifactorials at ca. 50% inbreeding potential and the great extent of factor specificities in both uni and bifactorials suggest some detrimental effect associated with a high degree of inbreeding

TABLE 1. The effect of the structure of the incompatibility factors on inbreeding potential

| | | | | Inbreeding |
|---|---|---|---|---|
| **Unifactorials** | | | | |
| A | | | | |
| A$\alpha$ | | A$\beta$ | | 50% |
| | | | | 75% |
| **Bifactorials** | | | | |
| A | | B | | |
| A$\alpha$ | A$\beta$ | B$\alpha$ | B$\beta$ | 25% |
| A$\alpha$ | A$\beta$ | B$\alpha$ | B$\beta$ | |
| A$\alpha$ | A$\beta$ | B$\alpha$ | B$\beta$ | 56% |

and a strong tendency for outbreeding. Furthermore, from laboratory experience with *Schizophyllum* a sharp decline in fruiting capacity is observed after 5-6 generations of inbreeding followed sometimes by a decline in basidiospore viability. The importance of outbreeding to *Collybia velutipes* was also indicated by Simchen (1965). The flexibility in the selection of the optimal level of inbreeding is the prerogative of the bifactorials only.

The outbreeding potential is determined solely by the number of factor specificities (Table 2). In the unifactorials with 4 factor specificities the outbreeding potential is 75%; with 20 specificities the outbreeding potential is 95%. A 2-locus structure would seem to be an improvement based on higher efficiency since with as few as 5 specificities at each locus 25 factor specificities would be formed and the outbreeding potential of such a population would be 96%. However, one cannot neglect the fact that the advantage of this organization would be offset by an immediate increase in the inbreeding potential, as stated above. It appears that the option of the unifactorials cannot be exploited if inbreeding at a level higher than 50% is detrimental to the species.

In the bifactorials the advantage in the ability to control the level of inbreeding along with a high outbreeding potential requires that the population maintain many factor specificities. The determination of factor specificity by 2 loci in each factor is the most efficient way to attain this goal, since only few mutations in an individual locus could provide numerous factor specificities. Furthermore, as indicated by Simchen (1967) the large number of factor specificities in the population can be maintained efficiently by very few strains.

In both the unifactorials and the bifactorials the number of factor specificities exceeds the number of factors required for securing close to 100% outbreeding efficiency. Whereas in the unifactorials this may reflect the nature of the locus with its maximum potential expressed, in the bifactorials it is tempting to speculate that it is a result of the evolution of the bifactorials from an ancestral unifactorial that already possessed some 20 allelic specificities. The evolution from the unifactorial to a typical bifactorial with 4 loci was in the early stage already an improvement

TABLE 2.  Organization of the incompatibility system:  Effect of factor structure and number of allelic specificities on outbreeding potential

| Type of Incompatibility system | Component loci | Number of alleles at each locus | | | | Outbreeding potential* (%) |
|---|---|---|---|---|---|---|
| | | Aα | Aβ | Bα | Bβ | |
| Unifactorial | Aα | 2 | – | – | – | 50 |
| | | 10 | – | – | – | 90 |
| | | 20 | – | – | – | 95 |
| | Aα,Aβ | 5 | 5 | – | – | 96 |
| | | 10 | 10 | – | – | 99 |
| Bifactorial | Aα,Bα | 5 | – | 5 | – | 64 |
| | | 20 | – | 20 | – | 90 |
| | Aα, Aβ: Bα,Bβ | 3 | 2 | 3 | 2 | 69 |
| | | 5 | 4 | 5 | 4 | 90 |

*For details of the calculations see Stamberg and Koltin (1973).

in the regulation of the inbreeding potential but with a severe effect on the outbreeding potential, thus creating strong selective pressure for the generation of new factor specificities at the new loci(us). A reinstatement of the very high level of outbreeding created the necessity to generate the new factor specificities by either a rapid accumulation of mutations at one new locus or the generation of the 2-loci structure of each factor and with it the accumulation of very few mutations at each locus. It is evident that the accumulation of even a few new specificities in the one new locus leads immediately to the formation of numerous specificities in one of the 2 incompatibility factors and the accumulation of very few mutations may already lead to a number of specificities that exceeds the level for maximum outbreeding potential (Stamberg and Koltin, 1973). As a corollary to this evolutionary consideration it is anticipated that an inequality in the number of factors in the $A$ and $B$ series will be found in distantly-related species of the higher fungi, and in cases where the allelic distribution is known, those will be asymmetrically distributed among the component loci. On theoretical grounds symmetrical evolution of alleles, assuming similar mutation rates in the component loci, is the most efficient evolutionary pathway for a rapid increase in the outbreeding potential (for the calculations see Stamberg and Koltin, 1973). The apparent contradiction supports this hypothesis for the mode of origin of 2-loci factors.

The number of alleles at each of the loci comprising the $A$ and $B$ factor is known only in Schizophyllum (Raper, Baxter and Ellingboe, 1960; Parag and Koltin, 1971; Stamberg and Koltin, 1973). The distribution of the alleles is clearly asymmetrical, with 32 alleles at $A\beta$ and 9 alleles each at the loci $A\alpha$, $B\alpha$ and $B\beta$. In at least 5 other species of the bifactorials, projected estimates are available based on factor repeats in the population (Table 3). In every species the number of $A$ and $B$ factors is unequal. This inequality suggests different number of alleles at the component loci or possibly a distribution similar to the one found in Schizophyllum. The occurrence of such an inequality in distantly-related species suggests that the asymmetry in Schizophyllum is not a result of an inherent property of the loci involved but rather a trend in the evolution of the incompatibility system. That in fact the loci are similar, may

TABLE 3.  Distribution of incompatibility factor specificities in bifactorials

| Species | A | B | Source |
|---|---|---|---|
| Schizophyllum commune | 288 (9 × 32)* | 81 (9 × 9) | Raper, Baxter and Ellinghoe (1960) Parag and Koltin (1971) Stamberg and Koltin (1973) |
| Coprinus lagopus | 400 | 100 | Estimated by Raper (1966) based on Hanna (1925) |
| Coprinus fimetarius | 400 | 100 | Estimated by Raper (1966) based on Brunswick (1924) |
| Polyporus obietinus | 70 | 200 | Estimated by Raper (1966) based on Fries and Jonasson (1941) |
| Pleurotus ostreatus | 63 | 190 | Eugenio and Anderson (1968) |

*The number of alleles at Aα, Aβ, Bα and Bβ, respectively.

even represent a series of duplications, is suggested by the similarity in mutational spectra of the various alleles examined both in $B\alpha$ and $B\beta$ (Raudoskoski, Stamberg, Bawnik and Koltin, 1976).

Each locus consists of two regions, a regulatory region involved in regulation of sexual morphogenesis and a recognition region involved in the discriminatory functions in the mating process. The structure of each locus bears a high degree of similarity to the structure of incompatibility genes of plants (Lewis, 1960) and similar to the genes of the immune system in mammals (Cooper and Lawton III, 1974).

The origin of the multiple specificities at each locus in the higher basidiomycetes and in plants, like the generation of antibody diversity in animals, remains to date an unsolved problem. All attempts to derive new alleles both in a higher fungus (Parag, 1962; Simchen, upublished; Raper, Boyd and Raper, 1965; Koltin and Raper, 1966; Koltin, 1968; Raper and Raper, 1973; Raudoskoski, Stamberg, Bawnik and Koltin, 1976) and in plants (Pandey, 1977) by mutagenesis have failed. All the mutations induced in natural alleles of Schizophyllum (including mutations recently isolated by Koltin, Stamberg, Bawnik and Tamarkin, unpublished) have led to the loss of specificity or loss of the regulatory function rather than to a new allelic specificity (Table 4). Even a two-step induction procedure, the induction of mutations in the primary mutants that had lost the allelic function, only succeeded in reverting the primary mutation but never succeeded in the derivation of an allele equivalent to the known natural alleles (Raper et al., 1965; Koltin and Raper, 1966; Raper and Raudoskoski, 1968; Raudoskoski, 1970).

Attempts to derive new alleles by other genetic methods such as intragenic recombination performed by J.R. and C.A. Raper, and by Koltin, have failed. Derivation of new alleles by inbreeding as suggested by Pandey (1970) for plants is currently being attempted (Bawnik, unpublished) but has not yielded thus far any new alleles.

The negative results from mutational studies and intragenic recombination led us to consider a different possibility. The results of recombinational studies with the world-wide collection of $B$ incompatibility factors of Schizophyllum are suggestive of the mode

TABLE 4.  Primary mutations in the incompatibility factors of _Schizophyllum commune_

| Alleles | | No. repeats of same mutations | Allelic specificity | Regulatory function | Mutagenic treatment[+] | Reference |
|---|---|---|---|---|---|---|
| **Aα** | **Aβ** | | | | | |
| 1 | | 0 | -- | -- | EMS, NA, NG, UV | Raper et al (1965) |
| | | 6 | lost | lost | EMS, NA | " |
| **Bα** | **Bβ** | | | | | |
| 1 | | 2 | unchanged | lost | X-ray | Koltin et al (unpub.) |
| 2 | | 1 | lost | lost | X-ray | Raudoskoski et al (1976) |
| 3 | | 1 | unchanged | lost | X-ray | Parag (1962) |
| | | 0 | -- | -- | Spontaneous | " |
| | | 0 | -- | -- | NM | Koltin (1968) |
| 6 | | 0 | -- | -- | X-ray | Koltin et al (unpub.) |
| 7 | | 3 | -- | -- | X-ray | Simchen (pers. com.) |
| | | 0 | -- | -- | Acriflavine | |
| 2 | | 1 | lost | lost | Spontaneous | Parag (1962) |
| | | 2 | lost | lost | NM | " |
| | | 4 | lost | lost | X-ray | Koltin (1968) |
| 4 | | 2 | lost | lost | X-ray | " |
| 5 | | 0 | -- | -- | X-ray | Koltin et al (unpub.) |
| 6 | | 1 | lost | lost | X-ray | Raudoskoski et al (1976) |
| 7 | | 2 | unchanged | lost | X-ray | Simchen (pers. com.) |
| | | 1 | lost | lost | Acriflavin | Koltin et al (unpub.) |
| | | 2 | lost | lost | X-ray | |
| 1[1] | | 1 | unchanged | lost | X-ray | Koltin (1968) |
| 7[1] | | 3 | unchanged | lost | X-ray | " |

[+]EMS – Ethyl-methane-sulphonate;  NA– Nitrous Acid;  NG– N-Methyl-N- Nitro-N-Nitrosoguanidine;
UV– Ultraviolet Irradiation;  NM– Nitrogen Mustard

of origin of certain alleles.

The recombinational characteristics of the $B$ factors divide the world-wide sample into 3 classes: *Class I* - factors in this class recombine among themselves with an average frequency of 2.3%; *class II* - factors in this class do not recombine among themselves nor with factors of any other class; *class III* - factors in this class recombine at a low frequency (0.13%) with factors of class I. Recombinants were never found in crosses within class III or in crosses with class II factors (Koltin and Raper, 1967; Koltin, Raper and Simchen, 1967; Koltin, 1969). The 3 classes were therefore referred to as "recombining" (class I), "non-recombining" (class II) and "low-recombining" (class III).

A functional analysis based on the mating interactions with specific mutations indicated the presence of 3 functionally distinct classes of $B$ factors, and this grouping of the factors coincided with the division based on the recombinational behaviour (Koltin and Raper, 1967a). These studies showed that the mating behaviour typical to each class is associated with specific alleles at the $B\beta$ locus of class II and specific $B\alpha$ alleles in class III. The differences in mating behaviour and the recombinational properties of the 3 classes could all be explained if class II and class III alleles were associated with deletions between the two component loci of the $B$ factor and extending into the region of the incompatibility genes (Parag and Koltin, 1977).

The notion that certain alleles are associated with deletions was further strengthened by the results of an attempt to synthesize the entire series of factors that can be theoretically obtained from the allelic series of the $B\alpha$ and $B\beta$ loci. Unexpectedly, this series could not be completed (Stamberg and Koltin, 1971) and 4 specific combinations of the 49 possible combinations of class I could not be formed. The result was not due to general suppression of recombination, nor to the genetic control of recombination, nor to structural rearrangements between the two loci of the $B$ factor. These alternatives were all experimentally ruled out. The 4 combinations that could not be derived involved 4 specific alleles, 2 at each of the component loci. These results once again could be rationalized if the alleled at $B\alpha$ locus were associated with deletions overlapping deletions that were associated with $B\beta$ alleles. This interpretation was further strengthened in crosses of $B$ factors with the entire series of class I $B\alpha$ alleles with strains carrying a mutation in the $B\beta$ locus derived by X-irradiation (Stamberg and Koltin, 1973). This mutation failed to recombine with 3 $B\alpha$ alleles, 2 of which were the alleles suspected earlier to be associated with deletions. The identification of an additional allele that fails to recombine with the mutant was suggested to be due to the larger extent of the deletion in the $B\beta$ locus of the mutant strain. The need to use the mutation to detect the third allele suggested that the deletion in this wildtype $B\alpha$ allele is smaller than that found in the two that were earlier identified in the recombinational studies with natural alleles. Thus, indications were obtained that of 7 natural alleles of class I, 3 specific alleles are associated with deletions. Together with the information on $B\alpha$ alleles of class III of the entire series of alleles of $B\alpha$ that consists of 9 wild-type alleles 5 seem to be associated with deletions.

The studies were later extended to include all available secondary mutations of the $B\beta$ locus (Raper, Boyd and Raper, 1965; Koltin and Raper, 1966; Raper and Raudoskoski, 1968; Raudoskoski, 1970; Raper and Raper, 1973), most of which were induced with X-ray. The results (Table 5) indicate that two of the mutations recombined with all $B\alpha$ alleles, two mutations recombined with all the $B\alpha$ alleles with the exception of $B\alpha 4$ two mutations recombined with all $B\alpha$ alleles with the exception of $B\alpha 4$ and $B\alpha 7$ and one mutant recombined with all the alleles other than $B\alpha 4$, $B\alpha 7$ and $B\alpha 5$ (Stamberg and Koltin, 1971; Stamberg and Koltin, 1974; Stamberg, Koltin and Tamarkin, in press). $B\alpha 4$ and $B\alpha 7$ are the class I alleles that were indentified earlier in recombination studies with wild $B\beta$ alleles as those alleles that may be associated with deletions, $B\alpha 5$ was also identified as a specific allele that may be associated with deletion in the first recombination study with a secondary mutation. Thus, these 3 alleles can be distinguished by recombination tests since it is an inherent property of these alleles. The same alleles from different locations maintain this specific property and factors experimentally synthesized also retain the property of selectively recombining.

Raper and Raper (1973) characterized the secondary mutations on the basis of their mating interactions and arranged them in order of increased impairment of the ability to interact with strains carrying wildtype $B\beta$ alleles, in matings in which the mates have identical $B\alpha$ alleles. Thus, the impairment of the $B\beta$ locus can be examined. The order of increasing impairment in $B\beta$ function corresponds precisely with the order suggested by the recombinational tests (Table 5). The mutations that display no difficulties in recombination with any $B\alpha$ allele also display no impairment in the $B\beta$ function and can dikaryotize a partner carrying any $B\beta$. The mutations that fail to recombine with one of the alleles already display some impairment of the $B\beta$ function and they cannot dikaryotize certain mates.

The recombinational and functional data are explainable if many of the wildtype alleles and the mutations are associated with deletions which extend from the alleles themselves to the region between $B\alpha$ and $B\beta$ (Figure 1). The degree to which the loci are deleted determine both the allelic specificity and the ability to interact with other alleles. The degree to which the deletion extends into the region between the $B\alpha$ and $B\beta$ would reflect on its ability to recombine with other alleles. It is known that various genes involved in functions such as nuclear migration and sexual morphogenesis map in the $B$ factor region (Koltin and Stamberg, 1972; Raper and Hoffman, 1974) and the deletion of a portion of this region could be reflected in the impairment of mating ability such as that displayed by the various mutant-$B$ strains.

At present it is not feasible by recombinational studies to test if any of the other alleles are also associated with deletions but the indications obtained in relation to 5 alleles of a total of 9 identified in each of the two loci in the $B$ factor of *Schizophyllum* strongly suggests that this may well be the situation with all other alleles all of which have been generated by a series of deletions.

TABLE 5. Recombinational and mating behaviour of mutant Bβ Bβ strains

| Bβ | Recombination with partner carrying Bα | | | | Mating interaction[a,b] with partner carrying Bα | | | | |
|---|---|---|---|---|---|---|---|---|---|
| | 4 | 7 | 5 | 1,2,3, 6 | 4 | 2 7 | 6 | 1,3 5 | $1^1$, $2^1$ |
| 2 | + | + | + | + | + | −[c] | + | + | + |
| 2(1) | + | + | + | + | + | + | + | + | + |
| 2(1-3) | + | + | + | + | + | + | + | + | + |
| 2(1-1) | + | + | + | + | + | + | + | + | + |
| 2(1-6) | − | + | + | + | − | + | + | + | + |
| 2(1-9) | − | + | + | + | − | − | − | + | + |
| 2(1-4) | − | − | + | + | − | − | − | + | + |
| 2(1-5) | − | − | + | + | − | − | − | − | + |
| 2(1-2) | − | − | − | + | − | − | − | − | + |
| α3(2)-β2(1-8) | − | − | − | − | − | − | − | − | − |

"+" indicates that recombination or mating interaction occurs;

"-" indicates non-occurrence.

[a] "Mating interaction" here refers to the ability to donate nuclei and dikaryotize the partner. Data taken from Raper and Raper (1973).

[b] We assume that the partners have a common Bα allele.

[c] This is the normal reaction. Any wildtype Bβ allele "recognizes" itself and does not interact when the partner carries an identical Bα and Bβ.

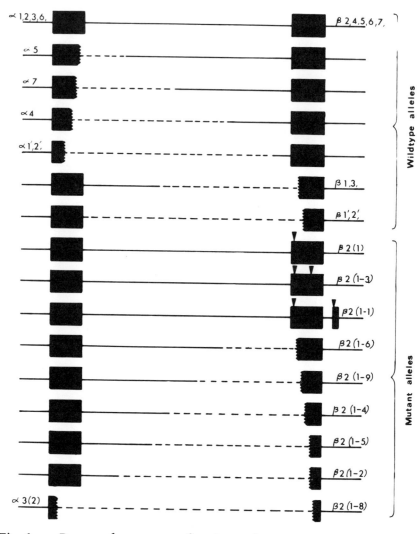

Fig. 1.    Proposed structure of native and mutant α and β loci of
the B incompatibility factor of *Schizophyllum*.  Dashed lines represent deleted chromosomal regions between the loci of the B
factor.  Indentation of the locus represents a deletion extanding
into the specificity region.  Arrows indicate point mutations.
(From Stamberg et. al., Molec. Gen. Genetics, inpress)

# REFERENCES

Berthet, P. and J. Boidin, 1966. Observation sur quelques hymenomycetes recoltes en Republique Cameronaise. Cah. Maboki 4: 27-54.

Boidin, J., 1966. Basidiomycetes Podoscyphaceae de la Republique Centrafricans, Cah. Maboki 4: 94-109.

Boidin, J. and P. Lanquentin, 1965. Heterobasidiomycetes saproohytes et homobasidiomycetes resupines. X. Novelles donnees sur la polarite dite sexualle. Rev. Mycol. (Paris) 30: 3-16.

Brunswick, H., 1924. Untersuchungen uber der Geschlechts - und Kemverhaltnisse bei der Hymenomyzetengattung, Coprinus. Bot. Abh. K. Goebel 5: 1-152.

Cooper, M.D. and A.R. Lawton III., 1974. The development of the immune system Sci. Amer. 231: 58-72.

Day, P.R., 1960. The structure of the A mating type locus in Coprinus lagopus. Genetics 45: 641-651.

Day, P.R., 1963. The structure of the A mating-type factor in Coprinus lagopus wild alleles. Genet. Res., Camb. 4: 55-65.

De Nettancourt, D., 1972. Self-incompatibility in basic and applied researches with higher plants. Genetica Agraria 26: 163-216.

Esser, K., 1967. Die Verbreitung der incompatibilitat bei Thallophyten. pp. 321-343. In W. Ruhland ed. Handb. Pflanzenphysiolo. 18.

Eugenio, C.P. and N.A. Anderson, 1968ı The genetics and cultivation of Pleurotus osteratus. Mycologia 60: 627-634.

Flexer, A.S., 1963. Bipolar incompatibility in Polyporus palustris. Thesis, Harvard University, Cambridge, Mass.

Fries, N. and L. Jonassen, 1941. Uber die interfertilitat verschiedener Stamme von Polyporus abietinus. Svensk. Bot. Tidskr. 35: 177-193.

Hanna, W.F., 1925. The problem of sex in Coprinus lagopus. Ann. Bot. 39: 431-457.

Jurand, M.K. and R.F.O. Kemp, 1973. An incompatibility system determined by three factors in a species of Psathyrella

(Basidiomycete). Genet. Res. Camb. 22: 125-134.

Koltin, Y., 1969. The structure of the incompatibility factors of *Schizophyllum commune:* class II factors. Molec. Gen. Genetics 103: 380-384.

Koltin, Y. and J.R. Raper, 1966. *Schizophyllum commune:* new mutations in the *B* incompatibility factor. Science 154: 380-384.

Koltin, Y. and J.R. Raper, 1967. The genetic structure of the incompatibility factors of *Schizophyllum commune:* three functionally distinct factors of *B* factors. Proc. Nat. Acad. Sci. U.S.A. 58: 1220-1226.

Koltin, Y. and J.R. Raper, 1967a. The genetic structure of the *B* incompatibility factors of *Schizophyllum commune:* the resolution of class III *B* factors. Molec. Gen. Genetics 100: 275-282.

Koltin, Y., J.R. Raper and G. Simchen, 1967. The genetic strucutre of the incompatibility factors of *Schizophyllum commune.* Proc. Nat. Acad. Sci. U.S.A. 57: 55-62.

Koltin, Y. and J. Stamberg, 1972. Suppression of a mutation disruptive to nuclear migration in *Schizophyllum* by a gene linked to the *B* incompatibility factor. J. Bacteriol. 109: 594-598.

Koltin, Y. and J. Stamberg, 1973. Genetic control of recombination in *Schizophyllum commmune:* Location of a gene controlling *B*-factor recombination. Genetics 74: 55-62.

Koltin, Y., J. Stamberg and R. Ronen, 1975. Meiosis as a source of spontaneous mutations in *Schizophyllum commune.* Mutation Res. 27: 319-325.

Lemke, P.A., 1969. A reevaluation of homothallism, heterothallism and the species concept in *Sistotrema brinkmannni.* Mycologia, 61: 54-66.

Lewis, D., 1960. Genetic control of specificity and activity of the S-antigen in plants. Proc. Roy. Soc. L., 151: 468-477.

Magni, G.E., 1964. Origin and nature of spontaneous mutations in meiotic organisms. J. Cellular Comp. Physiol. 64: Suppl. 1, 165-172.

Magni, G.E. and R.C. von Borstel, 1962. Different rates of

spontaneous mutations during mitosis and meiosis in yeast. Genetics **47**: 1097-1108.

Martin, J.M. and R.L. Gilbertson, 1973. The mating system and some other cultural aspects of *Veluticeps berkeleyi.* Mycologia **65**: 548-557.

Neuhauser, K.S., and R.C. Gilbertson, 1971. Some aspects of bipolar heterothallism in *Fomes cajanderi.* Mycologia **63**: 722-735

Pandey, K.K., 1970. New self-incompatibility alleles produced through inbreeding. Nature **227**: 689-690.

Pandey, K.K., 1977. Generation of multiple genetic specificities: Origin of genetic polymorphism through gene regulation. Theor. Appl. Genet. **49**: 85-93.

Papazian, H.P., 1951. The incompatibility factors and a related gene in *Schizophyllum commune.* Genetics **36**: 441-459.

Parag, Y., 1962. Mutations in the *B* incompatibility factor of *Schizophyllum commune.* Proc. Nat. Acad. Sci. U.S.A. **48**: 743-750.

Parag, Y. and Y. Koltin, 1971. The structure of the incompatibility factors of *Schizophyllum commune:* Constitution of the three classes of the *B* factors. Molec. Gen. Genetics **112**: 43-48.

Raper, C.A., 1976. Sexuality and life-cycle of the edible, wild *Agaricus bitorquis.* J. Gen. Microbiol. **95**: 54-66;

Raper, J.R., 1966. Genetics of Sexuality in Higher Fungi. Ronald Press, New York.

Raper, C.A. and J.R. Raper, 1973. Mutational analysis of a regulatory gene for morphogenesis in *Schizophyllum.* Proc. Nat. Acad. Sci., U.S.A. **70**: 1427-1431.

Raper, J.R. and R. Hoffman, 1974. *Schizophyllum commune.* Handbook of Genetics, Vol. 1 (R.C. King, ed.) pp. 597-626. Plenum Press, New York.

Raper, J.R. and M. Raudoskoski, 1968. Secondary mutations at the *B* incompatibility locus of *Schizophyllum.* Heredity **23**: 109-117.

Raper, J.R., M.G. Baxter and A.H. Ellingboe, 1960. The genetic structure of the incompatibility factors of *Schizophyllum commune:* the *A* factor. Proc. Nat. Acad. Sci., U.S.A. **46**:

833-842.

Raper, J.R., M.G. Baxter and R.B. Middleton, 1958. The genetic structure of the incompatibility factors in *Schizophyllum commune:* Proc. Nat. Acad. Sci., U.S.A. 44: 889-900.

Raper, J.R., D.H. Boyd and C.A. Raper, 1965. Primary and secondary mutations at the incompatibility loci in *Schizophyllum*. Proc. Nat. Acad. Sci., U.S.A. 53: 1324-1332.

Raudoskoski, M., 1970. A new secondary *B* mutation in *Schizophyllum* revealing functional differences in wild *B* alleles. Hereditas 64: 259-266.

Raudoskoski, M., J. Stamberg, N. Bawnik and Y. Koltin, 1976. Mutational analysis of natural alleles at the *B* incompatibility factor of *Schizophyllum commune:* $\alpha 2$ and $\beta 6$. Genetics 83: 507-516.

Roxon, J.E. and S.C. Jong, 1977. Sexuality of an edible mushroom *Pleurotus saju-caju*. Mycologia 69: 203-205.

Simchen, G. 1965. Variation in a dikaryotic population of *Collybia velutipes*. Genetics 51: 709-721.

Simchen, G., 1967. Genetic control of recombination and the incompatibility system in *Schizophyllum commune:* Genet. Res. Camb. 9: 195-210.

Stamberg, J., 1969. Genetic control of recombination in *Schizophyllum commune:* The occurrence and significance of natural variation. Heredity 25: 41-52.

Stamberg, J. and Y. Koltin, 1971. Selectively recombining *B* factors of *Schizophyllum commune*. Molec. Gen. Genetics 113: 157-165.

Stamberg, J. and Y. Koltin, 1973. The organization of the incompatibility factors in higher fungi: The effect of structure and symmetry on the breeding. Heredity 30: 15-26.

Stamberg, J. and Y. Koltin, 1973a. Genetic control of recombination in *Schizophyllum commune:* evidence for a new type of regulatory site. Genet. Res. Camb. 22: 101-111.

Stamberg, J. and Y. Koltin, 1973b. The origin of specific incompatibility alleles: a deletion hypothesis. Amer. Nat. 107: 35-45.

Stamberg, J. and Y. Koltin, 1974. Recombinational analysis of mutations at an incompatibility locus of *Schizophyllum*. Molec. Gen. Genetics **135**: 45-50.

Stamberg, J., Y. Koltin and A. Tamarkin. Deletion mapping of wildtype and mutant alleles at the *B* incompatibility factor of *Schizophyllum*. Molec. Gen. Genetics (in press).

Stimpfling, J.H., 1971. Recombination within a histocompatibility locus. Ann. Rev. Genet. **5**: 121-142.

Takemaru, T., 1961. Genetic studies on fungi X. The mating system in Hymenomycetes and its genetical mechanism. Biol. J. Okayama Univ. **7**: 133-211.

Takemaru, T. and N. Fujioka, 1969. The mating system of *Hirschioporus fuscoviolaceus* (Fr.) Donk. Rept. Tottori Mycol. Inst. (Japan) **7**: 59-63.

Takemaru, T. and N. Fujioka, 1970. Tetrapolar heterothallism in the basidiomycete *Steccherinum ochraceum* (Fr.) S.F. Gray. Rept. Tottori Mycol. Inst. (Japan) **8**: 27-32.

Takemaru, T. and S. Ide, 1970. Tetrapolar mating system in the basidiomycete *Microporus flabelliformis* (Fr.) Kuntze. Rept. Tottori Mycol. Inst. (Japan) **8**: 22-26.

Takemaru, T. and S. Ide, 1971. Tetrapolar heterothallism in the basidiomycete *Tyromyces pubescens* (Fr.) Imaz. Rept. Tottori Mycol. Inst. (Japan) **9**: 16-20.

Takemaru, T. and K. Oh'Hara, 1970. Incompatibility factors in the natural population of *Panus rudis* Fr. growing on a falling trunk. Rept. Tottori Mycol. Inst. (Japan) **8**: 33-38.

Terakawa, H., 1960. The incompatibility factors in *Pleurotus ostreatus*. Scientific Papers of the College of General Education Univ. Tokyo **10**: 65-71.

Ullrich, R.C., 1973. Sexuality, incompatibility, and intersterility in the biology of the *Sistotrema brinkmannii* aggregate. Mycologia **65**: 1234-1249.

Ullrich, R.C. and J.R. Raper, 1974. Number and distribution of bipolar incompatibility factors in *Sistotrema brinkmanii*. Amer. Nat. **108**: 507-518.

Welden, A.L. and J.W. Bennett, 1973. The cultural characterizations and matings type behaviour in *Podoscypha multizonata* and *P. ravenelii*. Mycologia **65**: 203-207.

Whitehouse, H.L.K., 1949. Multiple allelomorph heterothallism in the fungi. New Phytol. 48: 212-244.

# STUDIES ON MEIOSIS AND RECOMBINATION IN BASIDIOMYCETES

Judith Stamberg*

Basidiomycetes, in addition to being interesting objects of study in themselves, also serve as convenient tools for the exploration of genetic phenomena shared by all eukaryotes. In recent years much interest among geneticists has been focused on meiosis and on genetic recombination, in an attempt to understand the molecular basis of these central genetic processes. Two basidiomycete species, *Coprinus lagopus* and *Schizophyllum commune,* have contributed significantly to what we know today about these processes.

*Coprinus* is well suited to such investigations because meiosis is naturally synchronous throughtout the fruit body. The elegant work of Lu and associates has combined microscopy, genetics and biochemistry to exploit the synchrony of this species. The first study of the time sequence of meiosis in a fungus was done in *Coprinus lagopus* (Raju and Lu, 1970). At 25°C the entire process, from prefusion to release of mature basidiospores, takes more than forty hours. Meiotic DNA synthesis occurs before nuclear fusion and occupies about eight hours (Lu and Jeng, 1975). Pachytene, which is the longest meiotic stage, is about five hours. At this stage the synaptinemal complex can be seen by electron microscopy (Lu, 1967). Spindle pole bodies are also apparent at this and later stages in electron microscope studies (Lu, 1967). They are associated with spindle microtubules; however, their form and function is a matter of some controversy and has been studied in several basidiomycetes (McLaughlin, 1973; Setliff, Hoch and Patton, 1974; Gull and Newsam, 1976). From late pachytene until the end of diplotene the chromosomes have a "lampbrush" appearance (Lu and Raju, 1970). This is the first report in the literature of fungal chromosomes having a lampbrush appearance, though it is a well-documented phenomenon of diplotene in amphibians and insects, where it has been correlated with amplification of gene activities (Callan, 1963).

* Microbiology Department, Faculty of Life Sciences,
Tel—Aviv University, Tel—Aviv, Israel.

The stages from metaphase I till formation of tetrads are very short, altogether occupying only two hours. Spores are mature by eight to ten hours after meiosis is completed.

The synchrony of meiosis in *Coprinus* makes possible an investigation of the moleculer events connected with recombination. Three stages have been identified in which temperature shocks affect recombination between two closely linked markers (Lu, 1970, 1974). Heat shock increases the recombination frequency if given at premeiotic S phase, at leptotene-zygotene, or at pachytene; cold shock increases the recombination frequency only if given at pachytene. The latter is believed to be the stage at which recombination occurs (Henderson, 1970). According to the "hybrid DNA model" for recombination (which is widely accepted in its general outline, although specific details are controversial; see Holliday, 1974; Hotchkiss, 1974; Whitehouse, 1974), single-strand nicks are made in the DNA by an endonuclease, following which, strands unwind. Hybrid or heteroduplex DNA is formed in local, limited regions by the pairing of two single strands from homologous chromosomes. Some limited degradation of unpaired regions and resynthesis may occur. Any genetic heterozygosity in the heteroduplex region, causing mispairing, may also be repaired.

Repair-type synthesis of a small fraction of the total cell DNA indeed occurs regularly at pachytene; this has been demonstrated in the lily (Stern and Hotta, 1973) and has been shown to be true also in *Coprinus* (Lu and Jeng, 1975). Lu and Chiu (1976) showed that heat shock did not affect the rate of pachytene repair synthesis (as measured by $P^{32}$ incorporation); cold shock at pachytene, however, prevented $P^{32}$ uptake till after the end of the cold treatment. The authors concluded that although both heat and cold shocks can increase the recombination frequency, their effects on the recombination process are different. They suggest that heat causes nicks in the DNA and that the greater the number of nicks present at pachytene, the greater the recombination frequency. Cold shocks at pachytene would delay repair synthesis, which would likewise increase the recombination frequency.

In *Schizophyllum commune* meiosis is not naturally synchronous, and even along the gill there is no detectable developmental gradient (Radu, Steinlauf and Koltin, 1974). The chromosomes are very small and only a limited number of meiotic stages can be identified. These features have hampered cytological studies of meiosis. The synaptinemal complex of *Schizophyllum* has been described (Volz, Heintz, Jersild and Niederpruem, 1968; Radu, Steinlauf and Koltin, 1974) and the haploid chromosome number is thought to be 8 (Radu, Steinlauf and Koltin, 1974), though there are conflicting opinions (Haapala and Nienstedt, 1976).

Current studies on *Schizophyllum* in our laboratory make use of hydroxyurea to synchronize meiosis (Carmi, Koltin and Stamberg, 1977; Carmi, Raudaskoski, Stamberg and Koltin, 1977). Hydroxyurea (HU) has been found to inhibit mitotic and meiotic DNA synthesis in a number of organisms (Timson, 1975). We have found that a concentration of 0.075 M HU inhibits meiosis in *Schizophyllum* completely and reversibly. Young fruit bodies were transferred to medium containing HU, and gills were removed and stained at hourly intervals. Within three hours the percentage of meiotic cells in prefusion dropped from 26% (the control value) to 17%; by eight hours prefusion cells were at a stable 6%. For fusion cells, the trend was the opposite. Within three hours fusion cells rose from 70% to 77% of the total number of meiotic cells; by eight hours fusion cells represented more than 90% of the sample. The meiotic stages from metaphase I remained essentially constant throughout; postmeiotic stages immediately dropped in frequency. Sporulation stopped within 30 minutes of the treatment. Our interpretation is that nuclei in prefusion at the start of the HU treatment continue to fusion and stop at this stage. Nuclei already in fusion do not advance, and postfusion stages are also "frozen". Spores that are already mature are released, but no new spores begin the maturation process.

After an 8-hour exposure to HU, fruit bodies were returned to HU-free medium. Some 41 hours later, prefusion and fusion stages had returned to their normal frequencies but later stages of meiosis still occurred more slowly due to a lingering effect of the HU. Sporulation occurred in three waves after the release from HU. The first wave of spores fell at four to ten hours after release from HU, and was characterized by sparsity in number, very low germination, and an exclusively uninucleate condition (normally, all spores are binucleate due to a mitotic division). The next wave of spores began at 19 hours after release from HU; this wave was more abundant and had normal germination but consisted of predominantly uninucleate spores. From 25 hours onward, sporulation rapidly returned to normal with respect to number; at 40 hours the percentage of binucleate spores began to rise till, by 49 hours after the release from HU, sporulation was normal in every respect.

The relative DNA contents of treated and untreated meiotic nuclei were determined by microspectrophotometry of Feulgen-stained fruit bodies. In the untreated control the amount of DNA in samples of prefusion, telophase II and spore nuclei was the same, with very little variation from nucleus to nucleus for any of these stages. A sample of fusion nuclei, however, exhibited a much greater heterogeneity as reflected in the range and standard deviation. Fusion nuclei from HU-treated fruit bodies were homogeneous in DNA content. These data indicate that, in *Schizophyllum,* meiotic DNA replication occurs after fusion of nuclei. Prefusion nuclei are unaffected by HU and continue to the fusion stage, where DNA synthesis normally occurs. The heterogeneity in DNA content at the fusion stage in untreated fruit bodies is due to some fusion nuclei being in early stages of DNA replication while others have completed the replication. Fusion cells from HU-treated fruit bodies are homogeneous because no DNA replication occurs.

The order of meiotic events in *Schizophyllum* (i.e., nuclear fusion followed by DNA synthesis and meiosis) seems to be unique. In *Coprinus* and other fungi where the meiotic sequence has been studied, DNA synthesis always precedes nuclear fusion (Rossen and Westergaard, 1966; Lu and Jeng, 1975; Iyengar, Deka, Kundu and Sen, 1977).

The waves of sporulation which occur after release of fruit bodies from HU can be tentatively correlated with specific periods in the meiotic process. The spores released in the first few minutes after the start of HU treatment would be those that were in the final postmeiotic maturation process. After release from HU, the first wave of spores (at 4-10 hours after release) would be those caught by HU during the postmeiotic, premitotic DNA replication. The next wave (at 19-25 hours) would contain the nuclei that were stopped by HU at the beginning of premeiotic DNA synthesis following fusion. The final sporulation, characterized by complete normalcy, would have nuclei that were in prefusion stages at the start of HU treatment.

Thus, the timing of meiotic events in *Schizopyllum* can be at least roughly reconstructed. Meiosis, beginning with nuclear fusion and initiation of meiotic S phase and ending with sporulation, occupies about 25 hours under these conditions. This period can be divided into: premeiotic S and early prophase, 6-10 hours (the times are not exact because samples were taken only at 3-hour intervals); late prophase, and completion of first and second meiotic divisions, about 10 hours; premitotic S, 6-10 hours. Mitosis and spore maturation would be very short stages. These are maximal values, because residual HU in the fruit bodies after completion of the treatment probably prolongs the stages considerably. The results correlate reasonably well with the meiotic time schedule of *Coprinus* (see above), and can be tentatively correlated with the recent finding of Bromberg and Schwalb (1976) that, in *Schizophyllum,* meiosis does not proceed in the dark past the "one-nucleus stage". Five hours after release from dark inhibition sporulation begins. It is not clear whether the light-requiring stage is pre- or post-S phase, but the timing described above suggests that the block to meiosis occurs after the S phase. This point needs clarification, however.

The release of synchronous populations of spores in *Schizophyllum* by the use of the HU technique is now being exploited for studies on the molecular events of recombination.

Another intriguing aspect of meiosis which is being studied in this synchronized system is the "meiotic effect". This term refers to the tendency for spontaneous mutation frequencies to be markedly higher after meiosis than after mitosis. First described in yeast (Magni and von Borstel, 1962), the meiotic effect has since been found in other fungi and is frequently associated with recombination of markers bracketing the mutated site. In *Schizophyllum* the meiotic effect has also been demonstrated, but no correlation with recombination was observed (Koltin, Stamberg and Ronen, 1975). Recent work indicates that an error-prone repair process may be responsible for the meiotic effect; a "mutator" strain with a high spontaneous mutation frequency was found to have very low endonuclease activity (Shneyour, Stamberg and Koltin, 1976). In bacteria, endonuclease-mediated DNA repaid is exact, in comparison with other repair systems (Witkin, 1976); the hypothesis is that in *Schizophyllum,* similarly, low endonuclease activity may allow expression of an error-prone process resulting in a high mutation frequency.

*Schizophyllum* has also figured prominently in studies on the genetic control of recombination. In the course of a survey on the distribution of incompatibility factor alleles in nature, Raper, Baxter and Ellingboe (1960) found that recombination between the subunit loci of the *A* factor varied from 3 to 23%. Similar natural variation in *B*-factor recombination was later found (Koltin, Raper and Simchen, 1967). The source of this variation has been clarified by means of inbreeding, backcrossing, and other genetic manipulations (Simchen, 1967; Stamberg, 1968; Simchen and Connolly, 1968; Stamberg, 1969a; Koltin and Stamberg. 1973). Recombination frequencies in the *A* and *B* factors are independently regulated by genes located outside of the *A* and *B* factors themselves. The controlling genes are loosely linked to the regions they control; one such gene, *B-rec-1,* has been precisely mapped and is located nine map units from *B* β. The alleles determining low recombination frequencies are dominant to those for high frequencies, and the former are found more frequently in nature than are the latter (Stamberg, 1969b). Gene systems have also been found which control frequencies of recombination in specific, limited regions both linked and unlinked to the incompatibility factors (Simchen and Stamberg, 1969a; Tang and Chang, 1974).

Simchen and Stamberg (1969b) proposed a model for the regulation of recombination in which there are two general types of control, coarse and fine. The coarse control consists of genes which together control the sequential steps involved in recombination throughout the genome. The lack or misfunction of any of these gene products would result in the absence of recombination throughout the genome. Mutants lacking a synaptinemal complex or an enzyme essential for recombination such as endonuclease or ligase are of this type. Although common enough in the laboratory, they are of striking rarity in nature.

The fine control is superimposed in the coarse, and consists of a number of genes which regulate the amount of recombination in specific, short chromosomal segments. The *B-rec-1* gene in *Schizophyllum* is of this type. Some specific recognition must be involved in the relationship between the controlling "*rec*" genes and the controlled segments, with any one controlling gene active at several segments spread throughout the genome. Each of the segments under the common control of a *rec* gene would possess a recognition site specific for the *rec* gene product. Stamberg and Koltin (1973) obtained genetic evidence for the existence of such recognition sites in the region between the $\alpha$ and $\beta$ subunits of the *B* factor of *Schizophyllum*. Possibly the recognition sites are places where DNA is preferentially nicked; *rec* gene products might block the access of endonucleases of these sites.

Thus, the gene systems controlling recombination frequency in the *A* factor, the *B* factor, and other regions in *Schizophyllum* are of the fine-control type. Only in *Neurospora* has a similar series of *rec* genes with local effects and "cog" (recognition) sites been precisely identified (Angel, Austin and Catcheside, 1970; Catcheside, 1974). However, variation in recombination frequencies is a feature very commonly found among natural populations of many types of eukaryotes, and presumably therefore it is of value. For the regions bounded by the *A* and *B* factors the biological importance of the control of recombination can be assessed. These factors are naturally occurring markers and the outcome of recombination is obvious: the higher the recombination frequencies in these regions, the higher the inbreeding potential. Stamberg (1969b) suggested that the two-locus incompatibility factor structure plus genes controlling recombination between the loci of a factor represents a compromise between the needs to maximize outbreeding (the two-locus structure increases the number of factor specificities and hence the outbreeding potential), and minimize inbreeding (by keeping the intrafactor recombination low). This view is supported by the dominance relations of the recombination genes, low being dominant to high, and by the greater prevalence in natural populations of the alleles for low recombination. Under special circumstances, such as colonization of a new area, or unfavorable conditions, high recombination would, however, be advantageous (Simchen, 1967) and could be selected for. There is evidence that temperature and other environmental agents affect recombination frequencies in *Schizophyllum* (Stamberg and Simchen, 1970) and many other organisms. Simchen and Stamberg (1969b) pointed out that the specificity of the fine controls for particular genetic regions ensures that every linkage relationship can be made to produce the optimal frequency of recombinants (the optimal frequency depending on the genes in the particular region of the genome, and on the environmental conditions). Thus, the recombination frequency between any two genes is not completely fixed by the physical distance between them; an important degree of flexibility is maintained by the presence of fine-control genes.

# REFERENCES

Angel, T., B. Austin and D.G. Catcheside, 1970. Regulation of recombination at the *his-3* locus in *Neurospora crassa*. Aust. J. Biol. Sci. **23**: 1229-1240.

Bromberg, S.K., and M.N. Schwalb, 1976. Studies on basidiospore development in *Schizophyllum commune*. J. Gen. Microbiol. **96**: 409-413.

Callan, H.G., 1963. The nature of lampbrush chromosomes. Intern. Rev. Cytol. **15**: 1-34.

Carmi, P., Y. Koltin and J. Stamberg. Meiosis in *Schizophyllum commune:* Premeiotic DNA replication and meiotic synchrony induced with hydroxyurea. (submitted for publication).

Carmi, P., M. Raudaskoski, J. Stamberg and Y. Koltin. Meiosis in *Schizophyllum commune:* The effect of hydroxyurea on basidiospore sporulation, germination, and nuclear number. Molec. Gen. Genet. (in press).

Catcheside, D.G., 1974. Fungal genetics. Ann. Rev. Genet. **8**: 279-300.

Gull, K. and R.J. Newsam, 1976. Meiosis in the basidiomycetous fungus, *Coprinus atramentarius*. Protoplasma **90**: 343-352.

Haapala, O.K., and I. Nienstedt, 1976. Chromosome ultrastructure in the basidiomycete fungus *Schizophyllum commune*. Hereditas **84**: 49-60.

Henderson, S.A., 1970. The time and place of meiotic crossing-over. Ann. Rev. Genet. **4**: 295-324.

Holliday, R., 1974. Molecular aspects of genetic exchange and gene conversion. XIII Intern. Cong. Genet. Genetics **78**: 273-287.

Hotchkiss, R.D., 1974. Molecular basis for genetic recombination. XIII Intern. Cong. Genet. Genetics **78**: 247-257.

Iyengar, G.A.S., P.C. Deka, S.C. Kundu, and S.K. Sen, 1977. DNA synthesis in course of meiotic development in *Neurospora crassa*. Genet. Res. (Camb.) **29**: 1-8.

Koltin, Y., J.R. Raper, and G. Simchen, 1967. The genetic structure of the incompatibility factors of *Schizophyllum commune:*

the *B* factor. Proc. Nat. Acad. Sci. (U.S.) **57**: 55-62.

Koltin, Y., and J. Stamberg, 1973. Genetic control of recombination in *Schizophyllum commune:* Location of a gene controlling *B*-factor recombination. Genetics **74**: 55-62.

Koltin, Y., J. Stamberg, and R. Ronen, 1975. Meiosis as a source of spontaneous mutations in *Schizophyllum commune.* Mut. Res. **27**: 319-325.

Lu, B.C., 1967. Meiosis in *Coprinus lagopus:* A comparative study with light and electron microscopy. J. Cell Sci. **2**: 529-536.

Lu, B.C., 1970. Genetic recombination in *Coprinus.* II. Its relations to the synaptinemal complex. J. Cell Sci. **6**: 669-678.

Lu, B.C., 1974. Genetic recombination in *Coprinus.* IV. A kinetic study of the temperature effect on recombination frequency. Genetics **78**: 661-677.

Lu, B.C. and S.M. Chiu, 1976. Genetic recombination in *Coprinus.* V. Repair synthesis of deoxyribonucleic acid and its relation to meiotic recombination. Molec. Gen. Genet. **147**: 121-127.

Lu, B.C and D.Y. Jeng, 1975. Meiosis in *Coprinus.* VII. The prekaryogamy S-phase and the postkaryogamy DNA replication in *C. lagopus.* J. Cell Sci. **17**: 461-470.

Lu, B.C. and N.B. Raju, 1970. Meiosis in *Coprinus.* II. Chromosome pairing and the lampbrush diplotene stage of meiotic prophase. Chromosoma **29**: 305-316.

McLaughlin, D.J., 1973. Ultrastructure of sterigma growth and basidiospore formation in *Coprinus* and *Boletus.* Canad. J. Bot. **51**: 145-150.

Magni, G.E., and R.C. von Borstel, 1962. Different rates of spontaneous mutation during mitosis and meiosis in yeast. Genetics **47**: 1097-1108.

Radu, M., R. Steinlauf, and Y. Koltin, 1974. Meiosis in *Schizophyllum commune.* Chromosomal behavior and the synaptinemal complex. Arch. Microbiol. **98**: 301-310.

Raju, N.B., and B.C. Lu, 1970. Meiosis in *Coprinus:* III. Timing of meiotic events in *C. lagopus* (sensu Buller). Canad. J. Bot. **48**: 2183-2186.

Raper, J.R., M.G. Baxter, and A.H. Ellingboe, 1960. The genetic structure of the incompatibility factors of *Schizophyllum commune:* the *A*-factor. Proc. Nat. Acad. Sci. (U.S.) **46**: 833-842.

Rossen, J.M., and M. Westergaard, 1966. Studies on the mechanism of crossing over. II. Meiosis and the time of meiotic chromosome replication in the Ascomycete *Neotiella rutilans* (Fr.) Dennis. C.R. Trav. Lab. Carlsberg **35**: 233-260.

Setliff, E.C., H.C. Hoch, and R.F. Patton, 1974. Studies on nuclear division in basidia of *Poria latemarginata.* Canad. J. Bot. **52**: 2323-2333.

Shneyour, Y., J. Stamberg, and Y. Koltin, 1976. Characterization of a strain with a high spontaneous mutation frequency in *Schizophyllum.* Genetics **83**: s 70-71.

Simchen, G., 1967. Genetic control of recombination and the incompatibility system in *Schizophyllum commune.* Genet. Res. (Camb.) **9**: 195-210.

Simchen, G., and V. Connolly, 1968. Changes in recombination following inbreeding in 'Schizophyllum. Genetics **58**: 319-326.

Simchen, G., and J. Stamberg, 1969a. Genetic control of recombination in *Schizophyllum commuune:* Specific and independent regulation of adjacent and non-adjacent chromosomal regions. Heredity **24**: 369-381.

Simchen, G., and J. Stamberg, 1969b. Fine and coarse controls of genetic recombination. Nature **222**: 329-332.

Stamberg, J., 1968. Two independent gene systems controlling recombination in *Schizophyllum commune.* Molec. Gen. Genet. **102**: 221-228.

Stamberg, J., 1969a. Genetic control of recombination in *Schizophyllum commune:* Separation of the controlled and controlling loci. Heredity **24**: 306-309.

Stamberg, J., 1969b. Genetic control of recombination in *Schizophyllum commune:* The occurrence and significance of natural variation. Heredity **24**: 361-368.

Stamberg, J., and Y. Koltin, 1973. Genetic control of recombination in *Schizophyllum commune:* evidence for a new type of regulatory site. Genet. Res. (Camb.) **22**: 101-111.

Stamberg, J., and G. Simchen, 1970. Specific effects of temperature on recombination in *Schizophyllum commune.* Heredity **25**: 41-52.

Stern, H., and Y. Hotta, 1973. Biochemical controls of meiosis. Ann. Rev. Genet. **7**: 37-66.

Tang, C.Y., and S.T. Chang, 1974. Variation in recombination frequencies in *Schizophyllum commune* and its genetic control. Aust. J. Biol. Sci. **27**: 103-110.

Timson, J., 1975. Hydroxyurea. Mut. Res. **32**: 115-132.

Volz, P.A., C. Heintz, R. Jersild, and D.J. Niederpruem, 1968. Synaptinemal complexes in *Schizophyllum commune.* J. Bact. **95**: 1476-1477.

Whitehouse, H.L.K., 1974. Advances in recombination research. XIII Intern. Cong. Genet. Genetics **78**: 237-245.

Witkin, E.M., 1976. Ultraviolet mutagenesis and inducible DNA repair in *Escherichia coli.* Bacteriol. Revs. **40**: 869-907.

# EVOLUTION OF INCOMPATIBILITY

Peter R. Day*

The opportunity to discuss the evolution of incompatibility puts me in something of a quandry. On the one hand the assignment would cause some mycologists to catalogue in detail all the incompatibility systems they know leaving their audience to appreciate the complexity and variety of what has evolved. On the other hand others might seize the chance to try to convince you that basidiomycetes evolved from a red alga, or that their particular theory about the mechanism of specifity determination is the only one that counts.

In some respects I am on safe ground if I chose to discuss why and how incompatibility systems have evolved since no one can prove me wrong. Of course I run the risk that some molecular biologist will duplicate their evolution in a test tube but, as far as I know, this is not a real threat at the moment. The basis for my speculation is the large body of work on Schizophyllum which the two previous speakers have covered. Some years ago I worked on incompatibility in a species of Coprinus and so have some first hand acquaintance with at least one basidiomycete.

In April of 1974 John Raper and I were corresponding in another context, that I will refer to later, and he enclosed "for perspective" a xerox copy of a comment by Roy Watling (1971).

"Few higher fungi have been used to study the genetics of fungi as a whole and it is rather unfortunate that those which have been chosen are not always typical of the group, e.g. Schizophyllum, or that there has been some confusion in the identity of the fungus used e.g. *Coprinus cinereus* (as *C. lagopus* Day, 1959)."

Watling certainly has a point. Simply because there has been more genetical work on Schizophyllum than all the other higher basidiomycetes put together this should not blind us to the fact that

* The Connecticut Agricultural Experiment Station
123 Huntington Street
New Haven, Connecticut 06504

it may not be typical or representative of the group as a whole.

More recent experience of my own with other fungi such as Ustilago and Endothia, which are not higher basidiomycetes, has emphasized this point.

As Esser has pointed out (see Esser and Blaich, 1973) there are two kinds of incompatibility *homogenic incompatibility*--the kind shown by Schizophyllum and Coprinus and *heterogenic incompatibility*. The latter is perhaps better known as vegetative incompatibility. In fungi it regulates the outcome of fusion or anastomosis between hyphae of genetically different mycelia.

In *Schizophyllum commune* and *Coprinus cinereus* heterogenic systems are unknown and yet they may be quite common in other higher basidiomycetes with unifactorial or bifactorial homogenic systems. In these forms vegetative incompatibility would appear as intersterility (Esser and Blaich, 1973). A recent example was described by Ullrich and Anderson (1977) who found six intersterile groups of *Armillaria mellea* from Vermont. In each group a series of alleles was found at each of the homogenic incompatibility loci. Comparisons of alleles between groups cannot be made because of complete intergroup sterility. There are other examples of similar sterility barriers --the geographic races of *Coprinus micaceus* (bifactorial) found by Vandendries and Robyn (1929), and intersterile forms of *Auricularia auricula* (unifactorial) (Barnett, 1937; Duncan and MacDonald, 1967) and *Fomes igniarius* (bifactorial) (Verrall, 1937). In *Agaricus bisporus* vegetative incompatibility has been reported between mushroom spawns of different genetic backgrounds (Atkey et al., 1973). Several other examples are discussed by Whitehouse (1949) and Esser and Blaich (1973).

In evolution sterility barriers, or isolating mechanisms, play the role of allowing particular adapted genotypes to emerge and predominate so that they are not returned to a genetic 'melting pot' at each sexual generation. Heterogenic incompatibility in the examples I have mentioned will function as an isolating mechanism in sexual reproduction. In all of these species the formation of a vegetative heterokaryon (dikaryon) has to precede sexual outcrossing. It will therefore lead to sympatric speciation. In the

Ascomycetes the role of heterogenic incompatibility appears to be different. In *Neurospora crassa* where recent work (Mylyk, 1976) has revealed at least 5 loci in addition to C, D, E described by Garnjobst (1953) and Garnjobst and Wilson (1956), and in homothallic *Aspergillus nidulans* (Butcher, 1968), vegetative incompatibility is not a barrier to mating or outcrossing. Although it may sometimes prevent mating in Podospora it does not always do so (Esser, 1971). What then is its function?

Some years ago I suggested that vegetative incompatibility might prevent or limit the risk of infection by suppressive or invasive cytoplasmic determinants (Day, 1968). Caten (1971, 1972) showed that vegetative incompatibility in *Aspergillus amstelodami* severely restricts the transmission of vegetative death--a cytoplasmic determinant that may well be a virus. My colleague Sandra Anagnostakis (in press) has shown that vegetative incompatibility interferes with the transmission of hypovirulence in *Endothia parasitica*.

In 1975, Hartl, Dempster & Brown suggested a theory for the adaptive significance of vegetative incompatibility. Although based on findings in Neurospora they proposed its validity for other fungi. Their premise was that "hyphal fusions if unchecked would permit the exploitation of an adapted type of mycelium by another type less well adapted to the environment in which they both grow. Exploitation would occur in the resultant heterokaryon by one nucleus dividing at the expense of the other. This would create a selection pressure for vegetative incompatibility".

A good example is provided by Neurospora heterokaryons with nuclei differing in the pair of alleles *I* and *i* reported by Pittenger & Brawner (1961). These heterokaryons are stable if the frequency of *I* nuclei is less than 30%. If *I* is above 30% its frequency rises to 100%. If the *I* nucleus carries a nutritional deficiency not supplied by the culture medium growth stops. The alleles *I* and *i* seem to have no other effect than their action in heterokaryons.

In the higher basidiomycetes exploitive situations of this kind that involve nuclei should be ruled out by the septal pore structure found in both homokaryons and dikaryons. Nuclear migration only occurs when either the pore structure, or the septum itself, is broken down as a result of interaction between nuclei carrying different alleles at loci controlling homogenic incompatibility (Giesy and Day, 1965). In Schizophyllum and *Coprinus cinereus* this is the B locus. In those species with dikaryons that undergo conjugate nuclear division resulting in rigorously binucleate cells, with or without clamp connections, there is a further protection against exploitive compatible nuclei that have breached the first barrier of the septal pore. I suggest that the pattern of free and unrestricted mating between genetically dissimilar strains that is implicit in heterogenic incompatibility may only have evolved with the safeguards provided by the basidiomycete septal pore.

Only in the basidiomycetes do we have such large and complex series of multiple alleles at one or two loci to promote outcrossing. In *Ustilago maydis,* where complex septal pores do not occur, cell fusion is controlled by only two alleles at one locus, called the *a* locus, as in nearly all other smuts, the series of more than 20 alleles at the second or *b* locus (Silva, 1972), which apparently control pathogenicity, has no effect in promoting outbreeding which is fixed at 50% by the two alleles at *a.* Other authors (Bandoni, 1973; Brough, 1974, Hung and Wells, 1975, and Flegel, 1976) have described a similar modified bifactorial incompatibility system in several genera of Tremellales. Why should a system of multiple alleles occur at one locus and not both? The addition of a second locus reduces inbreeding 50% but unless multiple alleles are present at both loci outbreeding cannot be greater than 50% irrespective of the number of alleles one of them may have. Perhaps Ustilago and the Tremellales are still evolving and have only produced multiple alleles at one of the loci. But I digress.

Although the septal pore caps will restrict passage of nuclei and the larger DNA containing orgnaelles such as mitochondria, the holes will probably allow passage of smaller determinants such as covalently closed circular DNA molecules (plasmids) and viruses. The risk of infection with suppressive cytoplasmic determinants and viruses could well result in problems when dikaryons occur as mosaics with many different components. Mosaics have been described from nature in the wood rotting fungi *Polyporus betulinus* and *Polystictus versicolor* by Burnett & Partington (1957) and they are likely to occur in other fungi in the absence of vegetative heterogenic incompatibility. In the presence of vegetative incompatibility such mosaics will be severely confined. Todd and Raymer (1977) have described narrow dark zones in decaying wood caused by *Coriolus versicolor, Stereum hirsutum, Phlebia merismoides* and *Hypholoma fasciculare.* These zones result from antagonistic interaction between adjacent mycelia belonging to the same species.

Are there other mechanisms that limit the spread of cytoplasmic determinants that do not at the same time compromise outcrossing? It seems that there are. Ross (1977) has reported that in *Coprinus congregatus* an infectious agent that disrupts meiosis can be obtained from, and transmitted to, homokaryons, but cannot be made to infect dikaryons. One explanation is that in the dikaryon the infectious agent is associated with one of the two nuclei. This is supported by the finding that di-mon matings between a dikaryon carrying the agent and healthy homokaryons showed that only one of the two kinds of resulting dikaryon showed the pale gill phenotype. This suggests that differentiation of the dikaryon confers protection against infection perhaps by preventing replication of the infecting agent. A similar cytoplasmically determined defect of meiosis was reported in *C. cinereus* (Day, 1959). Many cytoplasmic determinants will of course be self-eliminating especially when they interfere with spore formation.

We have considered the two roles of heterogenic incompatibility as a means of controlling exploitation and as a sexual isolating mechanism. I have proposed that the basidiomycete septal pore was a prerequisite for the evolution of a freewheeling, unrestricted, system of homogenic incompatibility that allowed these fungi to do without heterogenic incompatibility thus avoiding the paradox of simultaneously promoting and restricting genetic exchange.

What can we say about the evolution of incompatibility systems. Surprisingly little. With regard to their mechanisms we are up against a fundamental problem of modern biology; how is specificity determined or how does self differ from non-self. The two incompatibility systems differ in polarity. In one the association of non-identical alleles causes cell death. In the other non-identity promotes or releases a morphogenetic sequence. In the first complexity has evolved in the direction of many loci rather than many alleles whereas in the second there are many alleles at only one or two loci. This difference may simply mean that there are many ways to program cell death by heteroallelic interaction but very few ways of promoting or releasing morphogenesis by the same means.

The sheer glamour of the complexity of homogenic bifactorial incompatibility exemplified by Schizophyllum has deservedly attracted much attention for the challenge it still offers of understanding how it works. I suggest there is an equal challenge in the other systems--and that it might provide some keys to the first. For example can we usefully study mutation at loci controlling heterogenic incompatibility and begin to identify the cell systems involved.

In the time remaining I want to concentrate on homogenic incompatibility. I could discuss the incidence of bifactorial and unifactorial incompatibility with homothallism in genera like Sistotrema (Ullrich, 1973). I could comment on the hypothesis that alleles of Schizophyllum arise by deletions (Stamberg & Koltin, 1973), or on John Raper's contention that it is all so damned complicated that it could only have arisen once in some primitive progenitor of the higher basidiomycetes. Instead I shall talk briefly about a report of a trifactorial system in the fungus *Psathyrella coprobia* (Jurand & Kemp 1973). The report described the mating interactions among 73 homokaryons. The origin of these spores was as follows:-

The original isolate was collected by Kemp from horse dung in April 1967 in Scotland and deposited in the herbarium of the Royal Botanic Garden, Edinburgh, after identification (RW5243). In December 1969 a dikaryotic mycelium was obtained from a mass plating of spores from the herbarium specimen. Fruit bodies were formed in culture and basidiospores were harvested, dried, and stored in silica gel. In November 1970, 73 homokaryons were established and their mating interactions revealed 8 mating type classes in roughly equal frequencies. These were assigned genotypes A1B1C1, A2B2C2, A1B2C2 etc. Mating tests were carried out by inoculating the homokaryons 1.5 cm. apart so that nuclear migration could be studied. Only homokaryons that were assigned different alleles at all 3 loci formed fast growing dikaryons with clamps as a result of bilateral nuclear migration. No unilateral nuclear migration was observed in this species (Kemp, personal communication). Matings between homokaryons with common-C factors produced fast growing mycelia but only along the junction line and these had some false clamps. Homokaryons with a common-B factor with or without common-C (A≠B=C= or C≠) gave slow growing mycelia with false clamps and these also were only formed at the junction line.

That issue of Genetical Research arrived in Cambridge and New Haven in November of 1973. Early in 1974, Red, that is John Raper, called to ask my opinion of the report. We agreed that the origin of the 73 homokaryons was not clear, and that no breeding test of second generation progeny had been carried out to confirm that the fruiting dikaryons were in fact C≠. In the absence of this test they could have been C= and hence a homogenic and heterogenic system could be operating together.

Red asked me to join him in writing a letter to the editor of Genetical Research pointing to the need for additional evidence to support the claim of something so novel. I should add that Eric Reeve, the editor, has an interest in fungal sexuality that dates back at least to a Genetical Society meeting in London November 1960. Some weeks before the meeting Penguin Books Ltd. had been allowed to publish for the first time in Britain an unexpurgated edition of Lady Chatterly's Lover by D. H. Lawrence. One of the speakers apologised for the lack of suitable four letter words he could use to help his audience understand his paper on tetrapolar sexuality. Other mycologists have had even less success in devising such words. Dr. Reeve was sympathetic but firm. He thought our letter censured his journal a little too strongly and asked us to wait the few months needed to make additional tests. The last of the matter was a letter that Red wrote in reply to Reeve, with copies to Kemp, dated 17 April 1974. It began; "I'm sorry that the note that Day and I submitted to GENETIC (sic) RESEARCH was taken as censure of the journal or its editor. This was not intended. GENETIC RESEARCH has proved in the past to be of such quality that it was somewhat surprising to find therein a paper of such definitiveness that under close examination could at best earn the Scottish verdict of *not proven.*"

There follow some pages amplifying the points made in the original letter and I would like to read you the last two paragraphs.

"I hope I don't appear too belligerent in this, but the sexuality in the higher fungi is complex to begin with, and new departures in this field such as "trifactorial incompatibility" should be accompanied by rigorous proof. This is most certainly not the case here."

"Should anyone, you or the authors, care to take a sporting chance on this, I'll wager a bottle of fine Scotch whiskey that when more complete analysis is made of this material, it will prove to be bifactorial -- or even unifactorial."

And there, so far as I know, the matter still stands, unresolved.

## Addendum

Dr. Esser's suggestion that the third locus controls fruiting may be correct but since A≠B≠C≠ has true clamps, whereas B= or C= have false clamps, it evidently also controls dikaryon morphology. Kemp (personal communication) has continued work on *P.coprobia* but has had unexpected difficulties in fruiting isolated dikaryons of the genotype A≠B≠C≠. However, he was able to isolate 25 tetrads from one fruit body produced on a dish with several homokaryons that were being tested for mating type. Mating type tests of the mycelia resulting from these tetrads gave the following;

| (22 tetrads) | | (2) | | (1) | |
|---|---|---|---|---|---|
| A1 B1 C1 | | A1 B2 C2 | | A1 B1 C1 | |
| A2 B2 C1 | | A1 B2 C2 | | A1 B1 C1 | |
| A2 B1 C2 | | A2 B1 C2 | | A2 B2 C1 | |
| A1 B2 C2 | | A2 B1 C2 | | A2 B2 C1 | |

Kemp noted that in the majority class no two strains of a tetrad were compatible, that is, formed vigorous mycelia with true clamps. Kemp also found that plates inoculated with either of the two groups of minority class tetrads fruited and gave rise to the four mating types found in the majority class among which there was not heterozygosity for C. The underlying assumption in Kemp's analysis is that A≠B≠C≠ is necessary for the bilateral formation of a fertile dikaryon. On this premise his results are difficult to explain. An alternative premise is that A≠B≠C= is required for bilateral formation of a dikaryon. On this basis the original 8 mating types must be assigned different C specificities to recognise that A≠B≠C= is the standard of compatibility.

Testerstocks

|        |          |          |
|--------|----------|----------|
| (i)    | A1 B1 C1 | A1 B1 C1 |
| (ii)   | A2 B2 C2 | A2 B2 C1 |
| (iii)  | A1 B1 C2 | A1 B1 C2 |
| (iv)   | A2 B2 C1 | A2 B2 C2 |
| (v)    | A1 B2 C1 | A1 B2 C1 |
| (vi)   | A2 B1 C2 | A2 B1 C1 |
| (vii)  | A1 B2 C2 | A1 B2 C2 |
| (viii) | A2 B1 C1 | A2 B1 C2 |

Reassigning mating types on the basis only of tester stock reactions we obtain for the 25 tetrads;

| (22 tetrads) | A1 B1 C1 | (2) | A1 B2 C2 | (1) | A1 B1 C1 |
|---|---|---|---|---|---|
| | A2 B2 C2 | | A1 B2 C2 | | A1 B1 C1 |
| | A2 B1 C1 | | A2 B1 C1 | | A2 B2 C2 |
| | A1 B2 C2 | | A2 B1 C1 | | A2 B2 C2 |

Again in each of the three classes of tetrad no pairs give rise to clamped dikaryons by bilateral migration. However in the majority class this is now due to B= rather than C≠.

An additional assumption is that although A≠B≠C= produces dikaryons with true clamps, only A≠B≠C≠ dikaryons, with false clamps, produce fruit bodies. Perhaps the C locus of *P. coprobia* is like the mating type locus of *Neurospora crassa* which has two alleles *A* and *a*. Homoallelism (A x A or a x a) is required for heterokaryon formation but heteroallelism (A x a) is required for sexual reproduction and at the same time prevents heterokaryon formation.

I thank Sandra Anagnostakis for helpful discussion and Roger Kemp for comments and permission to present his unpublished information.

# REFERENCES

Anagnostakis, S. L. 1977. Vegetative incompatibility in *Endothia parasitica. Exper. Mycol.* (in press).

Atkey, P. T., R. J. Barton, M. Hollings and O. M. Stone. 1973. Compatibility between (mushrooms) spawn strains, and virus transmission. Annl. Rep. Glass. Crops Res. Inst. p. 120.

Bandoni, R. J. 1963. Conjugation in *Tremella mesenterica. Can. J. Bot.* 41:467-474.

Barnett, H. L. 1937. Studies in the sexuality of the Heterobasidiae. *Mycologia* 29: 626-649.

Brough, S. G. 1974. *Tremella globospora*, in the field and in culture. *Can. J. Bot.* 52:1853-1859.

Burnett, J. H. and M. Partington. 1957. Spatial distribution of fungal mating type factors. *Proc. Roy. Phys. Soc. Edinb.*, 26:61-68.

Butcher, A. C. 1968. The relationship between sexual outcrossing and heterokaryon incompatibility in *Aspergillus nidulans. Heredity,* 23:443-452.

Caten, C. E. 1971. Heterokaryon incompatibility in imperfect species of Aspergillus. *Heredity* 26:299-312.

Caten, C. E. 1972. Vegetative incompatibility and cytoplasmic infection fungi. *J. Gen. Microbiol.* 72:221-229.

Day, P. R. 1959. A cytoplasmically controlled abnormality of the tetrads of *Coprinus lagopuss. Heredity* 13:81-87.

Day, P. R. 1968. The significance of genetic mechanisms in soil fungi. *In* T. A. Toussoun, R. V. Bega, and P. E. Nelson (ed.) *Root Diseases and Soil-Borne Pathogens.* U. Cal. Press, Berkeley; 69-74.

Duncan, E. G. and J. A. Macdonald. 1967. Micro-evolution in *Auricularia auricula. Mycologia* 59:803-818.

Esser, K. 1971. Breeding systems in fungi and their significance for genetic recombination. *Molec. Gen. Genet.* 110:86-100.

Esser, K. and R. Blaich. 1973. Heterogenic incompatibility in plants and animals. *Adv. Genet.* 17:107-145.

Flegel, T. W. 1976. Conjugation and growth of *Sirobasidium magnum* in laboratory culture. *Can. J. Bot.* 54:411-418.

Garnjobst, L. 1953. Genetic control of heterokaryosis in *Neurospora crasssa. Amer. J. Bot.* 40:607-614.

Garnjobst, L. and J. F. Wilson. 1956. Heterocaryosis and protoplasmic incompatibility in *Neurospora crassa. Proc. Nat. Acad. Sci. U.S.A.* 42:613-618.

Giesy, R. M. and P. R. Day. 1965. The septal pores of *Coprinus lagopus* (Fr.) sensu Buller in relation to nuclear migration. *Amer. J. Bot.* 52:287-294.

Hortl, D. L., E. R. Dempster and S. W. Brown. 1975. Adoptive significance of vegetative incompatibility in *Neurospora crasssa. Genetics* 81:553-569.

Hung, C-Y. & K. Wells. 1975. Genetic control of compatibility in *Myxarium nucleatum. Mycologia* 67:1181-1187.

Jurand, M. K. and R. F. O. Kemp. 1973. An incompatibility system determined by three factors in a species of *Psathyrella* (Basidiomycetes). *Genet. Res.* 22:125-134.

Mylyk, O. M. 1976. Heteromorphism for heterokaryon incompatibility genes in natural populations of *Neurospora crassa. Genetics* 83:275-284.

Pittenger, T. H. and T. G. Brawner. 1961. Genetic control of nuclear selection in Neurospora heterokaryons. *Genetics* 46:1645-1663.

Ross, I. K. 1977. An infectious disorder for meiosis in Coprinus. *Second Int. Mycol. Congr. Abstr.;* 557.

Silva, J. 1972. Alleles at the b incompatibility locus in Polish and North American populations of *Ustilago maydis* (DC) Corda. *Physiol. Plant Pathol.* 2:333-337.

Stamberg, J. and Y. Koltin, 1973. The origin of specific incompatibility alleles: a deletion hypothesis. *Amer. Nat.* 107:35-45.

Todd, N. K. and A. D. M. Raymer. 1977. Genetic studies on intraspecific antagonism in natural populations of wood-decaying basidiomycetes. *Heredity* (in press) abstr.

Ullrich, R. C. 1973. Sexuality, incompatibility, and intersterility in the biology of the *Sistotrema brinkmannii aggregate. Mycologiaa* 65:1234-1249.

Ullrich, R. C. and J. B. Anderson, 1977. The species concept and natural populations of *Armillaria mellea*. *Second Int. Mycol. Congr.* Abstr.;689.

Vandendries, R. and G. Robyn. 1929. Nouvelles recherches experimentales sur le comportement sexuel de *Coprinus micaceus*. Deuxieme Partie. *Mem. Acad. R. Belg. Cl. Sci.* 9:117pp.

Verrall, A. F. 1937. Variation in *Fomes igniarius* (L). Gill. *Tech. Bull. Minn. Agr. Exp. Sta.* 117:41pp.

Watling, R. 1971. Basidiomycetes: homobasidiomycetidae. *Methods in Microbiol.* 4:219-236.

Whitehouse, H. L. K. 1949. Multiple-allelomorph heterothallism in the fungi. *New Phytol.* 48:212-244.

# INCOMPATIBILITY FACTORS
# AND THE CONTROL OF BIOCHEMICAL PROCESSES

## J.G.H. Wessels*

## The incompatibility system as a model
## for the regulation of cell differentiation

As has been outlined earlier in this Symposium, the incompatibility system of basidiomycetes such as *Schizophyllum commune* appears to represent a case of exposed regulatory genes (the incompatibility factors) that control other genes the expression of which eventually results in the formation of distinct cell types. In summary: The presence in a homokaryon of one wild-type allele of each of the two loci, $\alpha$ and $\beta$, of both the *A*- and the *B* factor results in a mycelium with monokaryotic morphology (uninucleate cells). For convenience this is termed the *A-off B-off* state of morphology. The presence of different alleles of one or both of the loci in a heterokaryon or the presence of a (primary) mutation in one of the loci in a homokaryon switches on the expression of distinct morphologies, depending on the factor affected. If the alterations concern both the *A*-factor and the *B*-factor, the dikaryotic morphology distinguished by binucleate cells and clamp connections (*A-on B-on*) ensues. If only the *B*-factor is involved, septal dissolution and nuclear migration - normally transitory in the formation of the dikaryon - becomes constitutive (*A-off B-on*). If only the *A*-factor is involved other processes related to dikaryon formation become operative but nuclear migration does not occur and fusions of hook cells with penultimate cells fail, resulting in a mycelium bearing pseudoclamps (*A-on B-off*). For convenience the latter two morphological states will often be referred to as *B-on* and *A-on*, respectively.

Although the incompatibility genes with their multiple alleles are clearly adapted to their role of controlling mating interactions, they may function in essentially the same way as regulatory elements

* Department of Developmental Plant Biology, Biological Centre, University of Groningen, Haren, Nederland

involved in cell differentiation during development of multicellular organisms. The conclusion that each locus is composed of a regulatory and a recognition subunit (cf. Koltin, this Symposium) and that the regulatory subunit contains information pertaining to individual parts of the controlled pathways (Raper & Raper, 1973) is very much in line with the structure of hypothetical integrator genes proposed by Georgiev (1969) and Britten & Davidson (1969).

If the incompatibility factors are considered as regulatory genes not essentially different from the integrator genes envisaged for development in higher eukaryotes, their unique character obviously relates to the recognition subunits that change the expression of these genes. Models have been proposed to explain how recognition of the products of different alleles might occur (Prevost, 1962; Raper, 1966; Kuhn & Parag, 1972; Ullrich, 1973), but none of these models have been put through biochemical tests. In fact it is still difficult to reconcile all properties of the system with known properties of either nucleic acids or proteins. However, the genetic evidence (Raper & Raper, 1973) at least gives strong indications that the products of the incompatibility alleles are inactive by themselves but are converted into active substances by interaction between the products of different alleles or by rare (primary) mutations in these alleles (positive regulation).

The special features of the recognition system, apart from posing a problem in its own right, can be used as a tool to influence the expression of the incompatibility genes, i. e. to effect the switch from one cell type to another. As mentioned above, such master switches have been proposed to act during spatial differentiation of cells in the development of multicellular organisms but their analysis proves difficult (Davidson & Britten, 1973). One reason is that mutations in such regulatory genes are often lethal because the viability of the developing organism as a whole very much depends on the proper interactions between different cell types. The advantage of systems like *S. commune* then would be that the different cell types grow independently of each other and that mutations that lock the switches in certain positions or otherwise affect the regulated part of the morphogenetic sequences can be studied in isolated cell types.

In discussing the relevance of studies on the incompatibility system for understanding cell differentiation in general, it is also tempting to relate some other properties of the system to phenomena central to developmental biology. For example, in the course of development cells may become determined to form a certain cell type but cell differentiation, - that is the appearance of cells with distinct morphological and functional properties -, may only occur after a number of divisions of the determined cells. When basidiospores of *S. commune* carrying primary mutations in the incompatibility genes are germinated, they do not immediately produce the (determined) cell type that is specified by these mutations (Koltin & Flexer, 1969; Koltin, 1970). Instead they first exhibit the monokaryotic morphology (*A-off B-off*) and only after a few days, allowing for many cell divisions do they express the expected phenotype: disrupted septa and an irregular nuclear distribution (*A-off B-on*) for *Bmut*-germlings and the dikaryotic morphology (*A-on B-on*) for *AmutBmut*-germlings. A similar phenomenon occurs when a piece or a macerate of a *Bmut*-mycelium is used to start a new culture (Marchant & Wessels, 1974). The mycelium grows as a monokaryon for a few days (2-3 days at 30$^{\circ}$C, 5-7 days at 25$^{\circ}$C) before the *B-on* phenotype is expressed. Because the transitions from monokaryotic to differentiated mycelium occurs rather synchronously, this phenomenon of temporal differentiation may also be of great value in a search for biochemical sequences that lead to the differentiated state.

Another phenomenon central to development of higher eukaryotes is the stability of the differentiated state. If differentiation in *S. commune* is brought about by associating different alleles of the incompatibility genes in a heterokaryon, any stability of the differentiated state can be tested by removing one of the nuclei and observing the morphology of the subsequently formed homokaryotic cells. Such experiments were already performed by Harder (1927) and his results have recently been confirmed and extended by examining the reversion of homokaryotic protoplasts obtained from heterokaryotic cells (Wessels et al., 1976). In general these results show that clamp initiation can continue for some time after dissociating the different nuclei of a dikaryon (cf. C.A. Raper, this Symposium).

The exploitation of a fungus like *S. commune* as a model system for development in eukaryotes may have the additional advantage that the kinetic complexity of the genome of fungi is only 2.5-11 times the complexity of the *E. coli* genome (cf. Dusenbery, 1975). On the other hand, the structure of the chromosomes of fungi appears quite similar to that of higher eukaryotes as evidenced by the presence of all the usual histones (Goff, 1976) and the presence of nucleosomes in chromatin (Morris, 1976; Noll, 1976). Information on the organization of the genome of *S. commune* is still scanty and will be summarized below.

Two approaches are possible to analyse the mechanisms by which the incompatibility genes switch on morphogenetic processes. One is to compare the total assembly of proteins or mRNAs of monokaryotic mycelium (*A-off, B-off*) with that of differentiated mycelia (*A-off B-on; A-on B-off; A-on B-on*). The magnitude of the differences found would relate to the extensiveness of the controlled system. An insight into what is primary in causing differentiation and what is secondary, that is the result of differentiation, would be gained by following changes as they occur during temporal differentiation in germinating basidiospores carrying mutations in the incompatibility genes. Another approach would be to search first for specific proteins that are causally related to the expression of the differentiated state. Once such proteins have been found, their regulation can be studied in detail along the lines indicated above.

Studies pertaining to both types of approaches will be reviewed. Since the search for specific enzymes involved in morphogenesis led to an investigation of septal dissolution, a brief description of the structure of hyphal wall and cross wall of *S. commune* will also be included.

### Organization of the genome of *S. commune*

Frankel and Ellingboe (1977) have recently constructed a linkage map indicating at least seven chromosomes in *S. commune*. Radu et al (1974) have derived a chromosome number of 8 for the haploid phase by staining basidial nuclei at meiosis. With the same techniques, but without reference to the former study, Haapala & Nienstedt (1976) reported n = 4. In our laboratory (N. Elmallah, unpublished) we have looked at mitotic chromosomes in protoplasts using both the Giemsa and Feulgen staining techniques. Eight chromosomes can be clearly distinguished for the haploid phase. A common-B diploid was shown to have 16 chromosomes and by cytophotometric measurements on Feulgen-stained protoplasts the nuclei were found to have twice the amount of DNA.

From buoyant densities in CsCl we estimated that the G+C contents of nDNA amd mtDNA from *S. commune* were 57.3% and 23.7%, respectively. The corresponding values calculated from melting curves were 57.8% and 19.6%, respectively. G+C values reported earlier by Villa and Storck (1968) on the basis of buoyant densities were somewhat higher, *viz.* 61% for DNA and 28% for mtDNA. Vanyushin et al. (1960) have reported a G+C content of 57.1 for total DNA.

On the basis of reassociation kinetics J.J.M. Dons (this laboratory, unpublished) estimated a kinetic complexity of $2.5 \times 10^{10}$ daltons for nDNA with *E. coli* DNA ($2.7 \times 10^9$ daltons) as a standard. Thus the nDNA of *S. commune* is approx. 9.3 times more complex than *E. coli* DNA and an amount (1C) of 0.042 pg per nucleus can be calculated. This value is twice that estimated by Haapala & Nienstedt (1976) on the basis of colorimetric determinations.


## Biochemical indices of differentiation

The large number of modifier mutations that disrupt part of the morphogenetic sequences in the *A-on* or *B-on* phenotypes suggests an extensive number of genes that are controlled by the incompatibility genes (Raper & Raper, 1966). A biochemical search for the extensiveness of the regulated system was initiated by Raper and Esser (1961) who showed a few antigenic differences of two compatible co-isogenic monokaryons as contrasted to the derived dikaryon. This study was extended by Wang and Raper (1969, 1970) who included *A*- and *B*-mutants and modified strains in their analysis comparing total protein and isozyme patterns obtained by polyacrylamide elecrtophoresis. The differences between the co-isogenic homokaryons and heterokaryons were quite large, e.g. the isozyme patterns of 14 out of 15 common enzymes were severely affected. Whether this gives a realistic impression of the extensiveness of the regulated protein batteries is doubtful, even if many of the differences are considered secondary or influenced by the physiological age of the cultures. It is likely that at least some of the patterns were artifacts of proteolytic digestion (cf. Pringle, 1975) or other protein dissociative forces. In an attempt to eliminate some of these difficulties we labeled the proteins of the strains to be compared with either $^3$H-leucine or $^{14}$C-leucine and prepared protein extracts from the mixed mycelia so that dissociative factors would influence the proteins of both strains in the same way (this laboratory, unpublished). After electrophoresis and counting the $^3$H and $^{14}$C label in slices of the gel it was possible to find more differences between e.g. a wild-type homokaryon and a co-isogenic B-mutant than between two co-isogenic wild-type homokaryons. However, the interpretation of the differences was difficult because the two isogenic wild-type homokaryons also displayed considerable differences, a result that agrees with that obtained for isogenic strains of Coprinus lagopus (Smythe & Anderson, 1971). Therefore, in our opinion the results with protein spectra do not yet give clear information on the number of different proteins affected during differentiation. However, the results of Wang and Raper (1969, 1970) did show that protein spectra as such were correlated with phenotypes rather than with the specific mutations or allelic interactions that brought about these phenotypes. A continuation of this type of approach with the techniques now available to analyze protein and mRNA populations should provide valuable information regarding the number of proteins regulated by the incompatibility genes and the level(s) at which this regulation operates.

Two studies have been done aimed at biochemical processes specifically related to the B-on phenotype. One of these concerns the process of septal dissolution that is constitutive in this phenotype and which will be discussed in some detail below. The other study started from the observation that mycelia with the A-off B-on phenotype grow much less efficiently than those with the A-off B-off wild type (Hoffman & Raper, 1971). On the basis of molar growth yields Hoffman & Raper (1974) calculated that a B-mutant strain grew about ten times less efficiently than an isogenic wild-type strain which had an ATP yield of 36 per mol glucose, close to the maximum theoretical figure. Isolated mitochondria from both strains, however, showed the same ATP production (P/O ratios - 2 for citrate) and the same ATPase activity. Since the coupling process in the mitochondria did not seem to be affected they considered a general uncoupling by a cytoplasmic ATPase as a possible cause for the lowered growth efficiency of the B-mutant.

Results from our laboratory suggest an alternative explanation for the low growth efficiency of the B-on phenotype. They show that the B-mutation affects the accumulation of some cell constituents much more than others (Table 1): particularly the accumulation of R-glucan, glycogen and triglycerides is lowered but the accumulation of protein, chitin and polar lipids is hardly affected. The type IV modifier when present in the B-mutant restores the B-off phenotype with some ultrastructural aberrancies remaining (Marchant & Wessels, 1973). Table 1 shows that it also restores the accumulation of cell constituents towards the wild-type level. It may be significant that extracts of mycelia with the B-on phenotype showed increased specific activities of R-glucanase, amylase and lipase but not of chitinase or protease while S-glucanase was not detected in any strain (Wessels, 1969a; Wessels & Niederpruem, 1967; this laboratory unpublished). This suggests, but does not prove, that in the B-on phenotype an increased degradation accompanying a normal or even higher rate of synthesis was responsible for the lowered accumulation of R-glucan, glycogen and triglycerides. Since these substances constitute a major fraction of the mycelium this would result in a waste of energy and a low molar growth yield.

In general the pattern of high activities of a number of hydrolytic enzymes in the growing A-off B-on mycelium is not unlike that in a A-on B-on mycelium during carbon starvation (Wessels & Niederpruem, 1967). In the latter case this leads to the net degradation of polymers in the preformed mycelium, the products being used for the construction of carpophores (Wessels,

1965). In the growing *B-on* mycelium, degradation would compete with synthesis of polymers and this is exactly what is seen at the morphological level. In such a mycelium normal septa are produced during divisions in the apical cells but as early as 1 hour after synthesis these septa are degraded again (Niederpruem, 1971).

Of all the morphogenetic processes involved in the formation of the dikaryon the degradative processes related to septal dissolution appear most amenable to biochemical analysis. Being a temporary process during dikaryon formation, its constitutive occurrence in the common-*A* heterokaryon and *B*-mutant strain is also accompanied by an apparent weakening of the hyphal walls. This suggests a general increase in lytic activity towards wall material, which should be detectable in extracts, rather than a precisely localized degradation process which would call for more sophisticated methods of analysis.

TABLE 1. Amounts of major cell constituents accumulated in cultures of mutant strains *A43/Bmut* and *A41/Bmut/MIV-11* relative to those accumulated by a co-isogenic wild-type strain (A41/B41) of *Schizophyllum commune.*

| | relative amount accumulated | |
|---|---|---|
| cell constituent | *A43/Bmut* | *A41/Bmut/MIV−11* |
| protein | 0.9 | 1.1 |
| chitin | 0.9 | 1.1 |
| S-glucan | 1.1 | 1.1 |
| R-glucan | 0.4 | 0.7 |
| glycogen | 0.3 | 0.4 |
| triglycerides | 0.3 | 0.6 |
| other lipids | 0.8 | 1.0 |

Data on polysaccharides and protein are from Wessels (1969a). Lipids were determined by growing the mycelia on $^{14}$C-glucose, extracting the lipids from dried mycelium with chloroform/methanol (2:1), washing the lipid extracts according to Folch, and scanning thin-layer chromatograms (White & Powell, 1966). The data refer to cultures harvested at the time that the glucose in the medium was just depleted.

## The hyphal wall and septum of *Schizophyllum commune*

Chemical analysis of wall components (Wessels et al., 1972; Sietsma & Wessels, 1976) combined with several ultrastructural techniques together with enzymatic dissection and cytochemical staining (Wessels et al., 1972; van der Valk et al., 1977) have revealed the composition and architecture of the longitudinal wall and the cross wall as summarized in Fig. 1. This model refers to the monokaryon only although, until now, no obvious differences have been detected in the dikaryon.

The surface of the hypha is covered with a mucilage that may also penetrate the compact wall layers. Different strains excrete various amounts of this mucilage in the medium (Niederpruem et al., 1977) and in some strains it is even absent, a useful property if enzymes from the medium have to be isolated (Wessels, 1969b). The mucilage is a glucan consisting of β-1,3-linked chains with branches of single glucose units attached by β-1,6 linkages on every third unit, on average, along the chain (Sietsma & Wessels, 1977).

The compact outer layer of the wall consists of the alkali-soluble S-glucan shown to contain only α-1,3 linkages (Wessels et al., 1972; de Vries, 1974; Sietsma & Wessels, 1977) although Siehr (1976) reported the presence of some α-1,6 linkages. We feel, however, that this may have resulted from cytoplasmic contamination of the wall preparation analysed. The S-glucan in the native wall as well as in precipitates from alkaline solutions has a typical X-ray diffraction pattern proving that it occurs in the wall in a microcrystalline condition. With negative staining irregular fibers with a distinct substructure can be discerned (van der Valk, 1976). It is possible that S-glucan is not only present as a rather homogeneous layer but that it also penetrates the underlying R-glucan/chitin layer.

At the outside of the S-glucan layer, freeze fractioning sometimes reveals a typical rodlet pattern. This pattern was originally thought to belong to the S-glucan layer (Wessels et al., 1972) but the analysis of a similar structure in microconidia of *Trichophyton mentagrophytes* (Hashimoto et al., 1976) suggests that it represents a thin but distinct layer of a glycoprotein which is very resistant to proteolytic digestion.

The inner layer of the wall, which is insoluble in alkali, represents the chemically most complex part of the wall. It contains the chitin microfibrils embedded in a matrix that consists mainly of glucan (R-glucan) but also contains hexosamines and amino acids linked in an as yet unknown way. The glucan has been shown to contain somewhat varying proportions of β1,3 linkages, Bβ1,6 linkages, β1,3,6 branching points and end groups, depending on the strain, and a macromolecular structure for the R-glucan of a monokaryon (strain 699) has been proposed (Sietsma & Wessels, 1977). Siehr (1976) also reported on the proportions of linkage types of R-glucan from the same strain but his ratios widely deviate from ours.

An interesting feature of the macromolecular structure of the R-glucan is that a major part of this highly branched glucan consists of chains that are identical to the branched glucan chains that constitute the mucilage (Sietsma & Wessels, 1977). This suggests that the production of mucilage may result from unbalanced synthesis of R-glucan component chains or from a defect in the assembly of these component chains. In accordance with this view recent experiments on the effect of carbon dioxide on carpophore formation in *S. commune* (Sietsma et al., 1977) show that carbon dioxide inhibits the synthesis of R-glucan and at the same time promotes the formation of mucliage. In addition, the R-glucan produced has properties which suggests that it contains fewer of the 'mucilage-like' chains.

The macromolecular structure proposed for R-glucan does not explain its extreme insolubility and therefore the existence of covalent linkages between chitin and the glucan, possibly involving hexosamines and amino acids, are involved. The fact that polyoxin D, a specific inhibitor of chitin synthetase, also inhibits the synthesis of R-glucan is regenerating protoplasts (de Vries & Wessels, 1975) supports this contention. In addition, recent experiments show that when chitin in the R-glucan/chitin complex is selectively degraded, the glucan part becomes soluble in alkali. Notwithstanding this intimate association, the ratio of R-glucan to chitin may vary in different parts of the wall. As depicted in Fig. 1., the chitin microfibrils are protruding through the R-glucan at the inside of the wall but at the outside of the inner wall layer they are completely covered with R-glucan. In this context it may be important to note that the ratio of linkage types and end groups and thus the macromolecular structure of R-glucan may vary widely in different regions of the wall and septum and that the structure proposed on the basis of chemical analysis only represents an average. Such differences may determine the susceptibility of the R-glucan to enzymic degradation (Sietsma et al., 1977).

A prominent feature of the cross wall is a central plate consisting of densely interwoven chitin microfibrils with little, if any, matrix of R-glucan. The chitin microfibrils run randomly except around the central pore, under the septal swelling, where they are oriented circularly. The chitinous septal plate is anchored in the R-glucan/chitin layer of the lateral wall with a ring that on cross section has a triangular appearance. On both sides the chitinous plate is covered with an R-glucan/chitin layer that is continuous with the R-glucan/chitin layer of the lateral wall. A thickening of R-glucan occurs in the septal swelling but the major part of this swelling is probably made up of a highly hydrated substance that may be similar to the mucilage at the outside of the hyphal wall. At least in the monokaryon, the septum appears devoid of S-glucan.

## Enzymatic degradation of the septum

A detailed description of dissolution of the septal apparatus in the growing B-on phenotype of S. commune (Marchant & Wessels, 1974) has shown that after synthesis of the complete septum the septal swelling is the first part to be degraded closely followed by the disorganisation of the parenthosomes which cover the septal swelling on both sides of the cross wall (not shown in Fig. 1). A progressive thinning of the cross wall then produces an enlargement of the central aperture sufficient to allow the passage of nuclei. During cross-wall dissolution cytoplasmic vesicles, possibly containing hydrolytic enzymes, appear to fuse with the plasmalemma at the site of degradation.

A variety of enzymes must be instrumental in the dissolution of the septal apparatus. Nothing is known about the chemical nature of the parenthosomes and thus the kind of enzymes degrading these structures remains elusive. The chemical nature of the cross walls,

**Fig. 1.** Model of the morphological and chemical structure of the lateral wall and the cross wall of *Schizophyllum commune.*

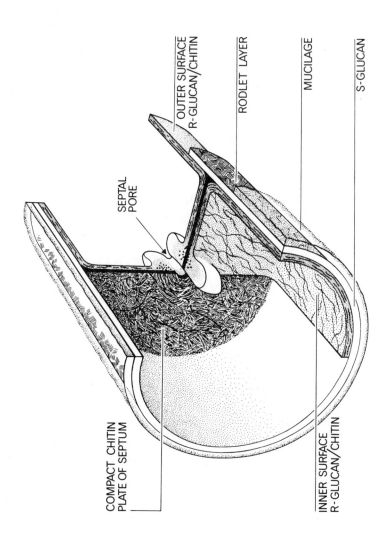

OUTER SURFACE
R-GLUCAN/CHITIN

RODLET LAYER

MUCILAGE

S-GLUCAN

SEPTAL
PORE

COMPACT CHITIN
PLATE OF SEPTUM

INNER SURFACE
R-GLUCAN/CHITIN

however, suggests that chitinase and β-glucanases are involved. With regard to glucanases that may attack R-glucan, *S. commune* produces an exo-laminarinase (β1,3-glucan glucohydrolase), an endolaminarinase (β1 3(4)-glucan glucanohydrolase), a pustulanase, and an enzyme called R-glucanase (Wessels 1969b). At the activity levels encountered in the organism, only R-glucanase, showed significant degradative activity towards R-glucan. Since the enzyme also hydrolyzed pustulan and produced large soluble glucan fragments it was tentatively given the systematic name β1,6-glucan glucanohydrolase but there also appears to exist an as yet unknown specificity requirement in addition to the presence of 6-substituted glucosyl units. The products of R-glucanase action on R-glucan are susceptible to laminarinases. Among these products, one bears a great similarity to the glucan chains of the mucilage and is particularly susceptible to exo-laminarinase.

In model experiments, isolated hyphal wall fragments containing cross walls were incubated with preparation of R-glucanase and chitinase after which they were examined both microscopically for septal dissolution and chemically for removal of wall components (Janszen & Wessels, 1970; Wessels & Marchant, 1974). It was found that the concerted action of both enzymes effected cross-wall dissolution while also removing R-glucan and chitin from the lateral wall. The R-glucan/chitin in the cross walls was more susceptible to enzymic degradation (as judged microscopically) than the R-glucan/chitin in the lateral walls (as determined by chemical analysis). A possible reason may be the presence of S-glucan in the lateral walls protecting R-glucan/chitin against degradation but differences in R-glucan structure may also play a role. After extensive degradation, however, only the S-glucan layer and the chitinous rings that anchored the cross walls in the lateral walls remained. With the electron microscope, cross-wall dissolution *in vitro* looked quite similar to that occuring *in vivo*.

In contrast to the cross walls of the monokaryon, the cross walls of the dikaryon proved much more resistant to chitinase and R-glucanase although the amounts of chitin and R-glucan solubilized from the whole preparation were similar to those solubilized from the walls of the monokaryon. This indicates that some difference

TABLE 2. Glucanase and chitinase activities in strains of *Schizophyllum commune* carrying mutations that affect septal dissolution.

| strain | septal dissolution | R-glucanase | chitinase | laminarinase |
|---|---|---|---|---|
| *A41/B41* | off | 14 (0.28) | 124 | 230 |
| *A41/Bmut* | on | 120 (3.21) | 92 | 551 |
| *A41/Bmut/MIV-11* | off | 30 (0.33) | 135 | 256 |

Isogenic strains carrying the specified mutations were grown for 4 days at 25°C in liquid surface cultures and enzyme activities were determined in the cytoplasmic fraction and in the medium (for R-glucanase only). Hyphae were broken in an X-press and R-glucanase and laminarinase activities were determined as described (Wessels, 1969a). Chitinase activity was determined after 24 h incubation with crustacean chitin and the products measured according to de Vries & Wessels (1973). Activities are expressed as ug monomer released per mg protein or ml medium (values in parenthesis).

Fig. 2. Relationship between genotype, specific R-glucanase activity in mycelial extracts, S-glucan/R-glucan ration in the wall, and morphology of the hyphae of 4-5 days old growing homokaryons and heterokaryons of *Schizophyllum commune*. *Bmut* and *Amut* refer to primary mutations in the Bβ2 and Aβ1 loci, respectively. *Bmut-mut* refers to a secondary mutation in the Bβ locus [β2(1-3)]. Data calculated from Wessels & Niederpruem (1967) and Wessels (1969a).

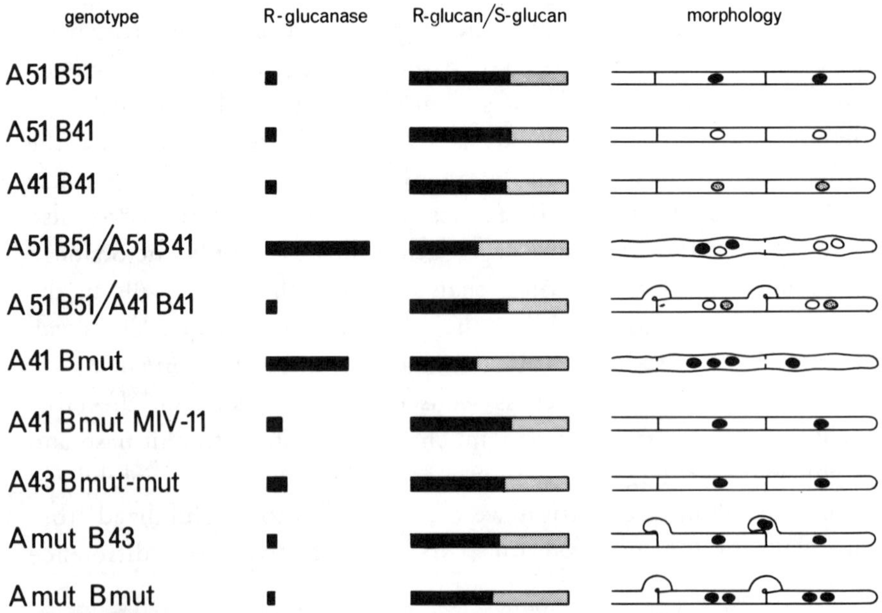

| genotype | R-glucanase | R-glucan/S-glucan | morphology |
|---|---|---|---|
| A51 B51 | | | |
| A51 B41 | | | |
| A41 B41 | | | |
| A51 B51/A51 B41 | | | |
| A51 B51/A41 B41 | | | |
| A41 Bmut | | | |
| A41 Bmut MIV-11 | | | |
| A43 Bmut-mut | | | |
| Amut B43 | | | |
| Amut Bmut | | | |

must exist between the cross walls of the monokaryon and the dikaryon, possibly in the structure of R-glucan of the cross walls. With the electron microscope, no difference was apparent. The difference in susceptibility may, however, be of major importance because it indicates that septal dissolution may be controlled both at the level of activities of degradative enzymes and at the level of susceptibility towards these degradative enzymes.

## Genetic regulation of septal dissolution

Table 2 shows that of three enzyme activities possibly involved in cross-wall dissolution only the R-glucanase activity is sharply increased in mycelia in which septal dissolution occurs (see also Wessels & Niederpruem, 1967; Wessels, 1969a; Wessels & Koltin, 1972). Laminarinases are also somewhat higher in the *B-on* phenotype but chitinase activities are not enhanced. Therefore R-glucanase activities were determined in various strains and heterokaryons and correlated with cell-wall composition and hyphal morphology. Figure 2 summarizes some of the results as obtained for growing 4-5 day old cultures, that is before the glucose in the medium was depleted. The salient fact is that, irrespective of the genotype, only the *B-on* phenotype with continous septal dissolution and irregularly shaped hyphae exhibits an increased R-glucanase activity and a low R-glucan content of the wall. This makes it likely that the causal relationship between R-glucanase activity and both septal dissolution and removal of R-glucan from the wall as determined in *in vitro* also holds *in vivo*. It also appears that one of the effects of the presence of different *A* alleles or of a mutated *A* allele is to suppress R-glucanase activity, thereby preventing septal dissolution. As mentioned above, another consequence of the operating *A* sequence may be the synthesis of septa that are less susceptible to R-glucanase.

Fig. 3 integrates some of the results in an attempt to visualize the R-glucanase/R-glucan interactions as they may occur under the direction of the incompatibility genes during the formation of a dikaryon after mating two compatible monokaryons. An as yet purely speculative feature of this scheme is that after fusion of the

**Fig. 3.** Model describing presumed changes in R-glucanase/R-glucan interactions during and after dikaryon formation. The R-glucan/chitin complex of the lateral wall and cross wall is indicated in black, the hatched area indicates dissolution of R-glucan. The dots represent R-glucanase activity. *AxBx* and *AyBy* represent any two monokaryons with different wild-type alleles at the *A*- and *B*-incompatibility factors. The cytoplasmic products specified by the alleles present in the acceptor strain (ax and bx) can only interact with those specified by the donor nuclei (ay and by) after the latter products have accumulated in the cytoplasm of the acceptor strain. The interaction of bx and by leads to an increase in R-glucanase and other proteins related to the *B-on* sequence. The interaction between ax and ay leads to switching on processes related to the *A*-sequence and to a blockage of part of the *B*-sequence, *viz.* septal dissolution, by suppressing R-glucanase and by effecting the synthesis of septa resistant to R-glucanase.

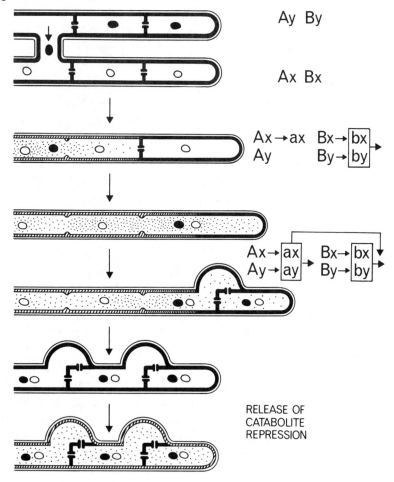

hyphae the temporal expression of events is based on a difference in the kinetics of formation of the cytoplasmic products of the $A$- and $B$ alleles, that is the accumulation of the product of the introduced $B$-allele in the foreign cytoplasm occurs faster than that of the introduced $A$-allele. Interaction of the products of different $B$-alleles would then activate the $B$-sequences without the $A$-sequences yet being operative. At this point, however, other mechanisms to explain temporal expression can also be envisaged (cf Holliday & Pugh, 1975). After hyphal fusion the $B$-on pathways would comprise the formation (or activation) of hydrolytic enzymes including R-glucanase which is instrumental in cross-wall dissolution. For R-glucanase this appears to be a general increase, not confined to certain domains in the hyphae, so that not only the septa but also the lateral walls are affected. However, the $B$-sequence also comprises localized processes at this stage such as the movement of nuclei through the widened apertures of the septal pores. Also, a high chitinase activity at the site of septal dissolution may be achieved by a directed transport of chitinase-containing vesicles that fuse with the plasmalemma around the dissolving septum. The scheme envisages that after a foreign nucleus has reached an apical cell, sufficient product of the foreign $A$-allele accumulates to interact with the product of the resident $A$-allele resulting in activitation of the $A$-sequence. One of the activities of the $A$-sequence is to interfere with the $B$-sequence by suppressing R-glucanase and by modifying newly synthesised septa in such a way that they are less susceptible to R-glucanase. In effect this activity of the $A$-sequence would switch off septal dissolution so that the other activities of this sequence can become expressed. This involves nuclear pairing, hook-cell formation, conjugate division of the paired nuclei and hook-cell septation. Only after the last event another part of the $B$-sequence, viz. hook-cell fusion, can become expressed. How this fusion process is accomplished at the molecular level is not known but it may be significant that this part of the $B$-sequence is also a lytic process.

Once the dikaryon is formed it grows with very low R-glucanase activity but this activity rises sharply again when the glucose in the medium is exhausted or experimentally removed (Wessels, 1966; Wessels & Niederpruem, 1967). Apparently the operation of the

*A*-sequence cannot suppress the R-glucanase activity when the system is releaved from catabolite repression. From this moment on the R-glucanase starts to degrade R-glucan in the walls of the preformed mycelium without much affecting the septa. As indicated by a number of experiments (cf. Wessels, 1965; Niederpruem & Wessels, 1969; Sietsma et al., 1977) the R-glucan then serves as a major reserve polysaccharide supporting the enlargement of the pilei of the carpophores. In this sense the formation of expanded pilei can also be regarded as part of the *B-on* morphogenetic sequence and this would explain why carpophores arising on *A-off B-off* mycelium (monokaryotic fruiting) mostly do not develop beyond the cup-shaped stage.

It is not yet possible to interpret the sequence of events as depicted above in terms of regulation of R-glucanase at the transcriptional or translational level. It is even uncertain whether R-glucanase refers to only one molecular species. Methods for specifically detecting R-glucanase protein(s) and mRNA are necessary to obtain clear answers regarding these point. In one instance at least R-glucanase appears to be regulated at the level of enzyme activity. In the double mutant *Bmut/MIV-11* (Fig. 3) the low activity of R-glucanase appears to be due to the presence of an inactive enzyme (Wessels & Koltin, 1972).

## Concluding remarks

The studies reviewed indicate the feasability of identifying molecular processes that both play a specific role in morphogenesis and are controlled by the incompatibility genes. Further research on the molecular mechanisms that control R-glucanase should reveal details of the mechanisms by which the incompatibility genes govern such processes. The identification of other proteins specifically related to morphogenesis is now of prime importance but it is doubtful whether enzyme measurements in extracts as done for R-glucanase will reveal such proteins. Many processes of the *A-on* sequence only occur at certain moments in the cell cycle and in specific domains of the cell. An example is the synthesis of septa resistant to enzymic degradation. For other processes such as hook-cell formation the molecular basis is still more obscure. More refined methods that can be used at the cellular level are needed to detect the molecular changes involved. On the other hand, the use of modern techniques for monitoring changes in populations of proteins or mRNA's could provide valuable complementary information regarding the extensiveness of the gene batteries that are controlled by the incompatibility genes, even if the function of individual gene products cannot as yet be identified. The incompatibility system of *S. commune* represents one of the few cases in eukaryotes in which the genetic basis of a regulatory complex involved in differentiation has been uncovered. This warrants further research on the molecular biology of the system.

# REFERENCES

Britten, R.J. & E.H. Davidson. 1969. Gene regulation for higher cells: a theory. Science 165, 349-357.

Davidson, E.M. & R.J. Britten. 1973. Organization, transcription, and regulation in the animal genome. Quart. Rev. Biol. 48, 565-613.

Dusenbery, R.L. 1975. Characterization of the genome of *Phycomyces blakesleeanus*. Biochim. Biophys. Acta 378, 363-377.

Frankel, C. and A.H. Ellingboe. 1977. New mutations and a 7-chromosome linkage maps of *Schizophyllum commune*. Genetics 85, 417-425.

Georgiev, G.P. 1969. On the structural organization of operon and the regulation of RNA synthesis in animal cells. J. theor. biol. 25, 473-490.

Goff, C.G. 1976. Histones of *Neurospora crassa*. J. Biol. Chem. 251, 4131-4138.

Haapala, O.K. & I. Nienstedt. 1976. Chromosome ultrastructure in the basidiomycete fungus *Schizophyllum commune*. Heriditas 84, 49-60.

Harder, R. 1927. Zur Frage nach der Rolle von Kern und Protoplasma in Zellgeschehen und bei der Ubertragung von Eigenschaften. Z. Bot. 19, 337-407.

Hashimoto, T, C.D. Wu-Yuan & H.J. Blumenthal. 1976. Isolation and characterization of the rodlet layer of *Trichophyton mentagrophytes* microconidial wall. J. Bacteriol. 127, 1543-1549.

Hoffman, R.M. and J.R. Raper. 1971. Genetic restriction of energy conservation in *Schizophyllum*. Science 171, 418-419.

Hoffman, R.M. & J.R. Raper. 1974. Genetic impairment of energy conservation in development of *Schizophyllum*: efficient mitochondria in energy-starved cells. J. Gen. Microbiol. 82, 67-75.

Holliday, R. and J.E. Pugh. 1975. DNA modification mechanisms and gene activity during development. Science 187, 226-232.

Janszen, F.H.A. & J.G.H. Wessels. 1970. Enzymic dissolution of hyphal septa in a basidiomycete. Antonie van Leeuwenhoek 36, 255-257.

Koltin, Y. 1970. Development of the *Amut Bmut* strain of *Schizophyllum commune*. Arch. Microbiol. 74, 123-128.

Koltin, Y. & A.S. Flexer. 1969. Alteration of nuclear distribution in *B*-mutant strains of *Schizophyllum commune*. J. Cell. Sci. 4, 739-749.

Kuhn, J. & Y. Parag. 1972. Protein-subunit aggregation model for self-incompatibility in higher fungi. J. theor. Biol. 35, 77-91.

Marchant, R. and J.G.H. Wessels. 1973. Septal structure in normal and modified strains affecting septal dissolution. Arch. Microbiol. 90, 35-45.

Marchant, R. and J.G.H. Wessels. 1974. An ultrastructural study of septal dissolution in *Schizophyllum commune*. Arch. Microbiol. 96, 175-182.

Niederpruem, D. J., Marshall, C., & J. L., Speth. 1977. Control of extracellular slime accumulation in monokaryons and resultant dikaryons of *Schizophyllum commune*.

Morris, N.R. 1976. Nucleosome structure in *Aspergillus nidulans*. Cell 8, 357-363.

Niederpruem, D.J. 1971. Kinetic studies of septum synthesis, erosion and nuclear migration in a growing B-mutant of *Schizophyllum commune*. Arch. Mikrobiol. 75, 189-196.

Niederpruem, D.J. and J.G.H. Wessels. 1969. Cytodifferentiation and morphogenesis in *Schizophyllum commune*. Bact. Revs. 33, 505-535.

Noll, M. 1976. Differences and similarities in chromatin structure of *Neurospora* and higher eukaryotes. Cell 82, 349-355.

Prevost, G. 1962. Etude genetique d'un Basidiomycete: *Coprinus radiatus* Fr. ex Bolt. Ann. Sci. Natur. Bot. 12th Ser. *3*, 425-613.

Pringle, J.R. 1975. Methods for avoiding artifacts in studies of enzymes and other proteins from yeasts. In: D.M. Prescott (ed.), Methods in Cell Biology XII, Academic Press, New York, pp 149-184.

Radu, M., R. Steinlauf & Y. Koltin. 1974. Meiosis in *Schizophyllum commmune*, chromosomal behaviour and synaptinemal complex. Arch. Microbiol. 98, 301-310.

Raper, C.A. & J.R. Raper. 1966. Mutations modifying sexual morphogenesis in *Schizophyllum commune*. Genetics **54**, 1151-1168.

Raper, C.A. & J.R. Raper. 1973. Mutational analysis of a regulatory gene for morphogenesis in *Schizophyllum*. Proc. Nat. Acad. Sci., USA 70, 1427-1431.

Raper, J.R. 1966. Genetics of Sexuality in Higher Fungi. The Ronald Press Co., New York.

Raper, J.R. & K. Esser. 1961. Antigenic differences due to the incompatibility factors in *Schizophyllum commune*. Z.Vererb. - Lehre 92, 439-444.

Raper, J.R. & C.A. Raper. 1973. Incompatibility factors: regulatory genes for sexual morphogenesis in higher fungi. Brookhaven Symp. Biol. 25, 19-38.

Siehr, D.J. 1976. Studies on the cell wall of *Schizophyllum commune*. Permethylation and enzymic hydrolysis. Can. J. Biochem. 54, 130-136.

Siestma, J.H., D. Rast & J.G.H. Wessels. 1977. The effect of carbon dioxide on fruiting and on degradation of a cell wall glucan in *Schizophyllum commune*. J. Gen. Microbiol. (in press).

Sietsma, J.H. & J.G.H. Wessels. 1977. Chemical analysis of the hyphal wall of *Schizophyllum commune*. Biochim. Biophys. Acta 496, 225-239.

Smythe, R. & G.E. Anderson. 1971. Electrophoretic protein spectra of wild-type and isogenic monokaryons of *Coprinus lagopus*. J. Gen. Microbiol. 66, 251-253.

Ullrich, R.C. 1973. Genetic determination of sexual diversity in the *Sistotrema brinkmannii* aggregate. Ph.D.thesis, Harvard University. 141 pp.

Valk, P. van der 1976. Light and electron microscopy of cell-wall regeneration by *Schizophyllum commune* protoplasts. Ph. D. Thesis, Groningen.

Valk, P. van der, R. Marchant & J.G.H. Wessels. 1977. Ultrastructural localization of polysaccharides in the wall and septum of the basidiomycete *Schizophyllum commune*. Exp. Mycol. 1, 69-82.

Vanyushin, B.F., A.N. Belozerskii and S.L. Bogdanova. 1960. A comparative study of the nucleotide composition of ribonucleic and deoxyribonucleic acids in some fungi and myxomycetes. Dokl. Akad. Nauk. SSSR. 134, 1222-1225.

Villa, V.D. & R. Storck. 1968. Nucleotide composition of nuclear and mitochondrial deoxyribonucleic acid of fungi. J. Bacteriol. 96, 184-190.

Vries, O.M.H. de 1974. Formation and cell wall regeneration of protoplasts from Schizophyllum commune. Ph.D. Thesis, Groningen.

Vries, O.M.H. & J.G.H. Wessels. 1973. Release of protoplasts from Schizophyllum commune by combined action of purified α-1,3-glucanase and chitinase derived from Trichoderma viride. J. Gen. Microbiol. 76, 319-330.

Vries, O.M.H. & J.G.H. Wessels. 1975. Chemical analysis of cell wall regeneration and reversion of protoplasts from Schizophyllum commune. Arch. Microbiol. 102, 209-218.

Wang, C.S. & J.R. Raper. 1969. Protein specificity and sexual morphogenesis in Schizophyllum commune. J. Bacteriol. 99, 291-297.

Wang, C.S. and J.R. Raper. 1970. Isozyme patterns and sexual morphogenesis in Schizophyllum. Proc. Nat. Acad. Sci. 66, 882-889.

Wessels, J.G.H. 1965. Morphogenesis and biochemical processes in Schizophyllum commune. Fr. Wentia 13, 1-113.

Wessels, J.G.H. 1966. Control of cell-wall glucan degradation during development in Schizophyllum commune. Antonie van Leeuwenhoek. 32, 341-355.

Wessels, J.G.H. 1969a. Biochemistry of sexual morphogenesis in Schizophyllum commune: Effect of mutations affecting the incompatibility system on cell-wall metabolism. J.Bacteriol. 98, 697-704.

Wessels, J.G.H. 1969b. A β-1,6-glucan glucanohydrolase involved in hydrolysis of cell-wall glucan in Schizophyllum commune. Biochim. Biophys. Acta 178, 191-193.

Wessels, J.G.H., H.L. Hoeksema & D. Stemerding. 1976. Reversion of protoplasts from dikaryotic mycelium of Schizophyllum *commune.* Protoplasma 89, 317-321.

Wessels, J.G.H. and Y. Koltin. 1972. R-glucanase activity and susceptibility of hyphal walls to degradation in mutants of *Schizophyllum* with disrupted nuclear migration. J. Gen. Microbiol. 71, 471-475.

Wessels, J.G.H., D.R. Kreger, R. Marchant, B.A. Regensburg & O.M.H. de Vries. 1972. Chemical and morphological characterization of the hyphal wall surface of the basidiomycete *Schizophyllum commune.* Biochim. Biophys. Acta 273, 346-358.

Wessels, J.G.H. and R. Marchant. 1974. Enzymic degradation of septa in hyphal wall preparations from a monokaryon and a dikaryon of *Schizophyllum commune.* J. Gen. Microbiol. 83, 359-368.

Wessels, J.G.H. and D.J. Niederpruem. 1967. Role of a cell-wall glucan-degrading enzyme in mating of *Schizophyllum commune.* J. Bacteriol. 94, 1594-1602.

White, H.B. & S.S. Powell. 1966. Fatty acid distribution in mycelial lipids of *Choanephora cucurbitarum.* Biochim. Biophys. Acta 116, 388-391.

# MORPHOGENTIC PROCESSES
# IN SCHIZOPHYLLUM AND COPRINUS

Donald J. Niederpruem*

"The student of morphogenesis in his haste
to use the methods of his colleagues in physiology
and biochemistry should not neglect this descriptive
part of his problem, for it may well point the
direction in which he must guide his steps."

E.W. Sinnott (1949)

To begin with, I would like to show you what the fruit bodies of *Schizophyllum commune* look like after you mate compatible monokaryons and incubate the petri dish in an upright position (Fig. 1). Dr. Walter Sundberg provided me with a lovely electron micrograph (Fig. 2) of a basidium sectioned through two basidiospores. One can clearly see the two nuclei in each spore. These photographs represent the questions we are addressing ourselves to: What controls the morphogenetic sequence which leads to structures such as these?

The approach in our laboratory has led to four areas of investigation which I want to review today.

1.  Cytology of basidial proliferation.
2.  Cytology of the initial events in fruiting.
3.  The control of polyol metabolism in morphogenesis.
4.  The control of extracellular slime production.

1.   The original experiments involved fitting a microscope with a ocular micrometer and watching live cells behave. We had already done this type of work with monokaryotic hyphae and we thought that if we examined live dikaryotic hyphae we could obtain an insight into the cytological events which underwrite basidiocarp formation. The first observation we made concerned the fact that

* Department of Microbiology, Indiana University School of Medicine
Indianapolis, Indiana  46202

there is a great deal of intercalation in the vegetative dikaryon behind the apex (Fig. 3). We already observed this phenomenon in spore germlings and monokaryotic hyphae. That is, the hyphae cut themselves up into smaller and smaller units as they grow. We were also excited about finding primary branches emerging from clamp connections about 6 or 7 cells from the dikaryotic apex Fig. 4). As you can see, as the primary branch emerges and forms a hyphal apex, nuclei can migrate into the new primary, initiate a new clamp connection after a mitosis and continue to grow apically (1). These data will help us to interpret the events in fruit body tissue. In Fig. 5 one can see a basidium of S. commune with a primary branch emerging from the clamp connection. Fig. 6 shows how successive outgrowths from the clamp connection can form new basidia. In this way, we believe the hymenium of S. commune proliferates and expands. The following list summarizes the various means of basidial proliferation found in fungi (2).

A.    Percurrent Proliferation
B.    Sub-basidial Lateral Proliferation
C.    Lateral Proliferation through Sub-basidial Clamp
D.    Apical Proliferation

In summary, our work suggests (3) that in S. commune the principal type of basidial proliferation is type (C) and the complete process is diagramed in Fig. 7. Percurrent proliferation would be indicated by intrahyphal hyphae and although we have never observed this in S. commune it is worth noting that hyphae can live inside of other hyphae in this species (4).

Another important yet poorly understood aspect of morphogenesis are the motive forces which operate within hyphal cytoplasim. Besides the well known issue of nuclear migration, vacuoles are also known to migrate apically. Fig. 8 demonstrates apical vacuole movement through a developing septum in S. commune. The sequence, which takes less than one-half hour shows that when a septum begins to form (by anular ingrowth) there are few vacuoles apically. However, as the septum starts to close, increasing numbers of vacuoles from the distal cells start to move forward. What is the role of these vacuoles and how do they move apically? We have only a few clues. Figure 9 shows a primary branch

Fig. 1. Fruit bodies of *S. commune*. Petri dish incubated in an upright position. 1x.

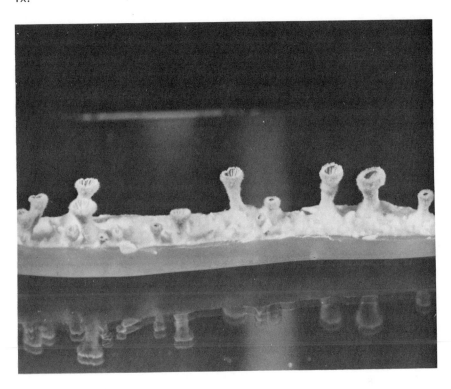

**Fig. 2.** Electron micrograph of a section through a basidium showing two of the basidiospores.

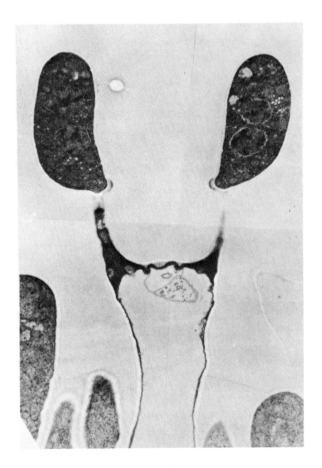

Fig. 3. Intercalation in the vegetative dikaryons of S. *commune.*

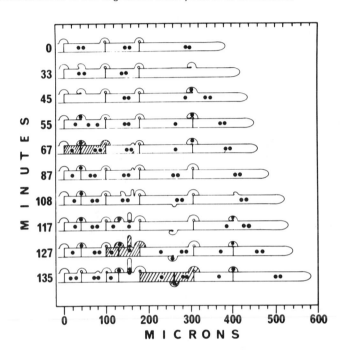

110

Donald J. Niederpruem

Fig. 4. Branch outgrowth through true clamp connections and subsequent clamp formation in vegetative dikaryons hypha of *S. commune*. (From CRC Critical Reviews in Microbiology, D. J. Niederpruem and R. J. Jersild, vol. 1, pg. 545-576, 1972. Used by permission of the Chemical Rubber Co.).

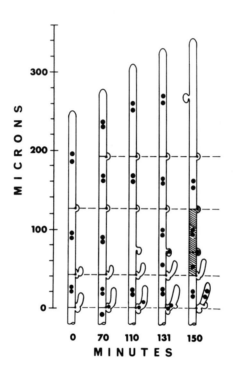

Fig. 5. Sub-basidial branching from clamp-connections in hymenium of dikaryotic fruit bodies. Nuclei migrating from penultimate cell into clamp outgrowth (x5800). (From ref. 3).

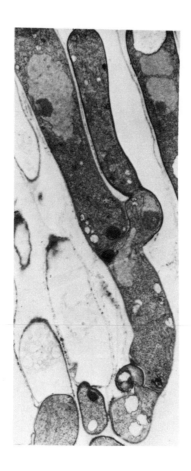

Fig. 6. Successive basidial proliferation through clamp connections in the hymenium of dikaryotic fruit body (From CRC Critical Reviews in Microbiology, D. J. Niederpruem and R. J. Jersild, vol. 1, pg. 545-576, 1972. Used by permission of the Chemical Rubber Co.).

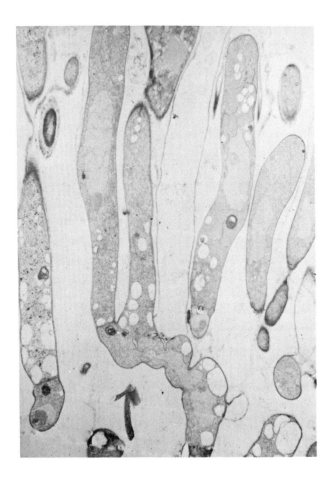

Fig. 7. Proposed mechanism of basidial proliferation in hymenium of dikaryotic fruit-body of *Schizophyllum commune*. A, binucleate basidium; B, diploidization and meiosis indicated by tripartrtite synaptinemal complexes. Probasidium emerges through clamp connection; C, four daughter nuclei occur from meiosis and four sterigmata occur on basidial apex. Penultimate nuclei enter presumptive basidium; D, four daughter nuclei enter basidiospore initials. Probasidium initiates clamp connection, undergoes mitosis and septal synthesis; E, mitosis in basidiospore produces binucleate condition; spore subsequently discharges, leaving collapsed basidium (shaded). New binucleate basidium now delimited while remaining daughter nuclei return to basal cell of origin. E and F, entire process is repeated. Figure diagrammatic and not to scale. (From ref. 3).

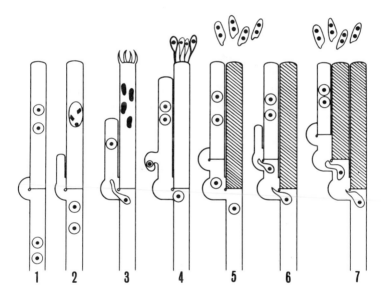

Fig. 8. Apical vacuolar movement through forming septum in hypha of *S. commune.* Time: A - 0 min; B - 5 min; C - 6 min; D - 8 min; E - 11 min; F - 17 min. (Unpublished observation of D. J. Niederpruem and P. L. Lane).

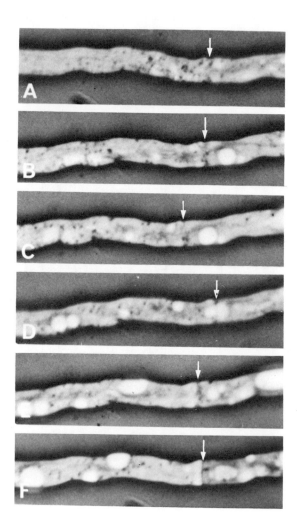

Fig. 9. Primary branch of vegetative hypha of *S. commune* occluded with vacuolar material. A - 0 min; B - 20 min; (Unpublished observation of D. J. Niederpruem).

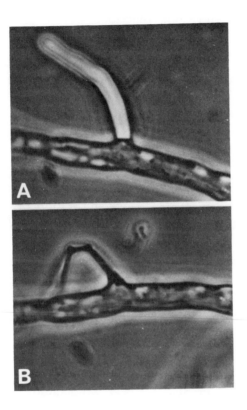

that has literally been gorged with vacuoles and just five minutes later the cell has burst. We also have observed many times that nuclei don't like to be within the reach of vacuoles and will do all they can to migrate away. While these observations may be coincidence, they may be indicative of vacuole function.

"Believe in him who seeks the truth:
But doubt him who finds it."

Montaigne

2.     Our technique for studying the initial events of fruit body morphogenesis in *Coprinus lagopus* was devised by Dr. Tom Matthews in our laboratory. He made an observation chamber in which a small block of nutrient medium with a cover slip on top was placed next to a block of water agar so that the dikaryotic mycelium starts growing on the nutrient block and is suddenly confronted with an area of nutrient depletion (5). We believed that this might act as a localized trigger for fruit body development and the sequence could then be followed. The results indicated that the first event is a localized initiation of primordia at the interface of the water agar block and cover slip (Fig. 10). Neither the cover slip nor the water-agar block will individually lead to localized fruiting. Furthermore, there is complete maturation of the fruit body at the point where the cover slip and water agar meet. This includes sporulation and autolysis. If we trace this process back we can observe the initial events. A lattice work builds up which becomes more intricate and complex. At some point we observe hyphal aggregates but we don't know if this is due to hyphal anastomosis or cohesion. There does seem to be cell interconnections.

When we section one of these tiny fluffs (initials) we can see the tangled mass of hyphae that represents the primordia a *C. lagopus* (Fig. 11). By the time the primordium is 0.5mm in diameter, apical differentiation is already evident. When the primordium is only 1.0mm in diameter (Fig. 12) it contains a "homunculus", a tiny, yet very complete mushroom (6). This mushroom already demonstrates biochemical differences between the base and the mushroom proper (Fig. 13) in that glycogen specifically accumulates at the base. We do not know the fate of that glycogen and if it is mobilized later in

Fig. 10. Localized fruiting in *C. lagopus* using the water-agar block technique (Unpublished observation of T. R. Matthews and D. J. Niederpruem).

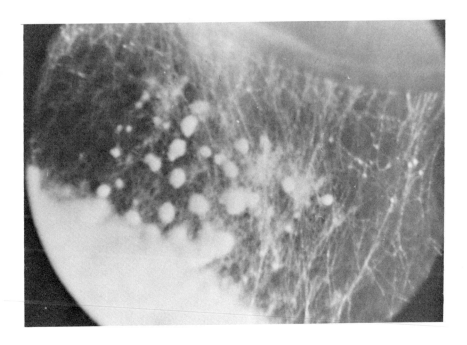

Fig. 11. Early stage of visible primordia from the hyphal lattice. X25. (From ref. 5).

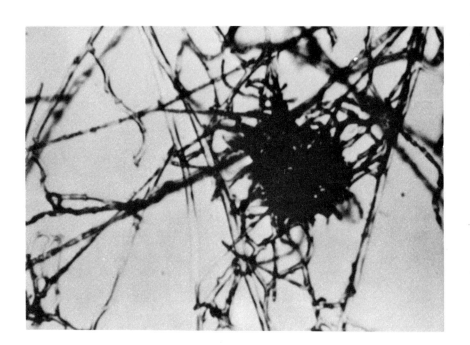

Fig. 12. Median vertical section of 1.0mm primordium of *C. lagopus* with well defined regions of the mushroom apparent. Methylene blue azure II - borax stain. (From ref. 6.).

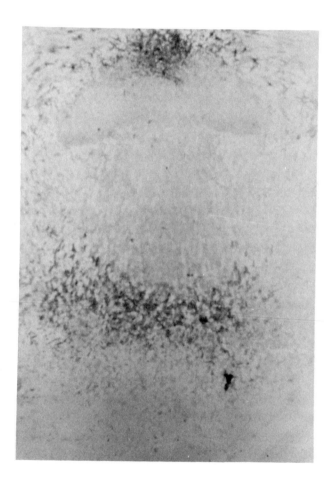

Fig. 13. PAS stain of bulbous basal cells of 1.0mm primordium. (From ref. 6).

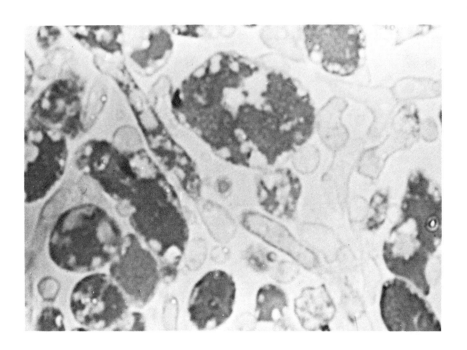

morphogenesis.

"No cell is an Iland
intire of it selfe"

3.     It is worth remembering at this point what Dr. Wessels told
me about eight years ago. He said that if you are not careful you are
going to have your head bitten off and I walked in anyway, where
angels fear to tread, and began to study some biochemical aspects of
morphogenesis.

We recognized early in the game that basidiospores contained
high levels of trehalose, mannitol and arabitol (7). During
glucose-dependent germination there is a depletion of polyols,
especially arabitol. There are variable, strain-dependent levels of
arabitol in monokaryotic hyphae. Dikaryons also have arabitol but
there is a conspicuous increase in the arabitol level in fruit bodies and
basidiospores (Fig. 14) (8).

It seemed obvious at this point that the next step should be a
study of the enzymes of polyol metabolism. Fig. 15 summarizes the
pathway of polyol metabolism in S. commune. The principal
reactions are the $NADH_2$-dependent reduction of fructose and
ribulose to form mannitol and arabitol, respectively. $NADPH_2$ will
not function as a coenzyme in these reactions. The difficulty in
understanding the relationship of enzyme activity to levels of polyols
is shown in Fig. 16. Here we observe high levels in strain 699 of all of
the enzymes associated with polyol metabolism even though no
arabitol can be found in this monokaryon.

We then examined the activities of these enzymes in a dikaryon
and its component monokaryons to determine if the activities might
be additive. The assay conditions were optimized for each strain and
the activities are never additive (Fig. 17) but rather the dikaryotic
activity was usually intermediate between the monokaryons.

We then examined the electrophoretic pattern of the polyol
dehydrogenase activity in two monokaryons, their dikaryon and the
other stages of morphogenesis (Fig. 18). Although different patterns
are found in each sample, no definitive relationship can be observed.
We were amazed to find eleven different arabitol dehydrogenase
"isozyme" bands in fruit body primordia, and I suggest that some of
these may be artifacts.

**Fig. 14.** Arabitol formation in Schizophyllum commune development on glucose medium. A basidiospores; B 5-h germlings; C 24 h through 5 day germlings; D homokaryon (str. 699), E homokaryon (str. 845), F dikaryon (699 x 845); G dikaryotic fruit-body (699 x 845). Culture ages examined were 2 day through 8 day for D, E and F while fruit bodies were from 9 day cultures. Cells are diagramatic and not drawn to scale. A spot size of 1 represents 10 ug of arabitol.

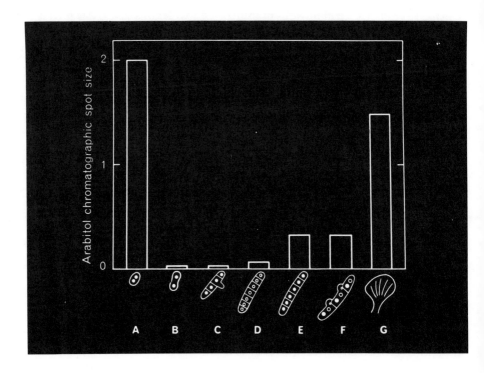

Fig. 15. Pathway of polyol metabolism in *S. commune*. (Unpublished observation of J. L. Speth & D. J. Niederpruem).

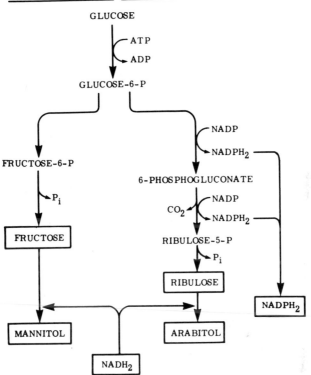

**Fig. 16.** Specific activities of enzymes of polyol metabolism in vegetative homokaryotic mycelium of *S. commune* comparing strains of 699 and 845 at various culture ages. (Unpublished observations of J. L. Speth and D. J. Niederpruem).

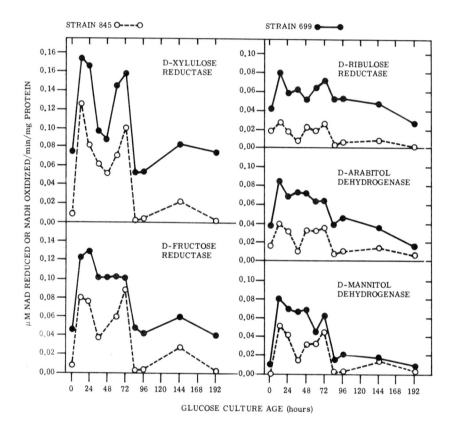

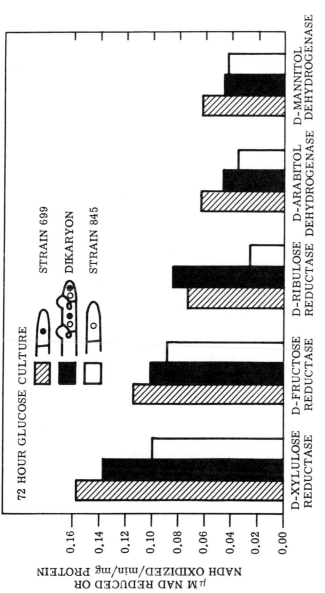

Fig. 17. Specific activities of enzymes of polyol metabolism in homokaryons in dikaryons of *S. commune*. (Unpublished observation of J. L. Speth and D. J. Niederpruem).

Fig. 18. Electrophoretic patterns of arabitol dehydrogenase(s) during morphogenesis of *S. commune.* (Unpublished observations of J. L. Speth and D. J. Niederpruem).

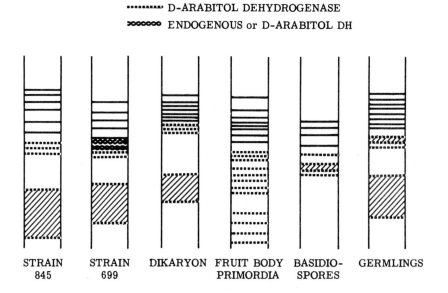

(Electrophoresis , 120min)

4.     I would now like to discuss some of our recent studies on the production of slime (mucilage) of *S. commune* and its regulation in morphogenesis. Dr. Wessels has already reviewed for you the structure and relationship of the slime to cell walls of *S. commune* . The slime is frequently produced in copious quantities and is easily measured by alcohol precipitation followed by a dry weight determination or anthrone reaction.

Fig. 19 shows the pattern of slime accumulation and dry weight in three different strains grown on glucose-asparagine medium (9). As you can see, the slime production lags behind the increase in dry weight during the first few days of growth. At about the 8th day, the slime production peaks and this is followed by several days of decline. By the 12th day the amount of slime in the culture levels off.

Recognizing that the reaching of the maximal level of slime after 8 days of growth appeared to be a general phenomenon, we then examined the slime accululation in 21 different monokaryotic strains of *S. commune* at 8 days (Table 1). The data shows that the quantity of slime produced is very dependent on the strain ranging from about equal the amount of dry weight of mycelium to no slime at all. We then decided to examine the genetics of slime production by mating strains with different capacities to produce slime. The results shown in Table 2 show that a dikaryon always produces significantly lower amounts of slime than the average of its constituent monokaryons (9). This is true even when both homokaryons are themselves high slime producers. If one looks at the kinetics of slime accumulation in two high slime producing monokaryons and their dikaryon (Fig. 20), the depression of slime synthesis in the dikaryon is clear.

It has been shown that slime is not required for fruiting (10) and when slime is present it is no longer synthesized after an early stage of fruiting (11). From the data present here it is now clear that slime production is decreased as a prelude to dikaryotic fruit body morphogenesis. In addition, we have shown in our lab that sealed cultures, which accumulate respiratory $CO_2$ show an *increase* in slime (9) and Dr. Wessels has told me that when $CO_2$ arrests fruiting, there is an increase in slime production (12). Finally, we also mated cohesiveless strains which do not fruit and these dikaryons show an *increase* in slime relative to their constituent monokaryons (9). These data suggest that studies of the enzymology of glucan synthesis and degradation in *S. commune* will be important areas of future research.

I would like to end with a picture of better times about five years ago.

Gerlind Eger, John R. Raper, Carlene A. Raper

"Through your students and your disciples
will come your greater honor"

Osler

Fig. 19. Effects of culture age on mycelial growth (as dry weight) and extracellular slime accumulation in three different monokaryons of *S. commune.* (Unpublished observations of J. L. Speth and D. J. Niederpruem).

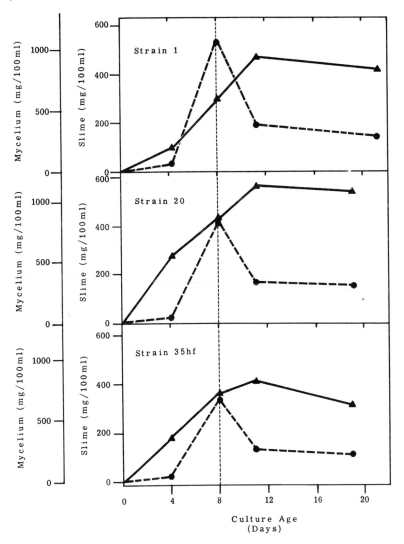

TABLE I:   Slime Accumulation and Vegetative
           Growth by Various Monokaryons of
           Schizophyllum commune After 8 Days
           of Culture

| Strain Number | Slime* | Mycelium* |
|---|---|---|
| 1 | 733 | 790 |
| R-MW46 | 700 | 775 |
| 5 | 575 | 605 |
| E-714 | 460 | 859 |
| R-848 | 400 | 917 |
| 20 | 390 | 542 |
| R-693 | 293 | 351 |
| 35hf | 250 | 597 |
| R-134 | 250 | 302 |
| E-605 | 237 | 1356 |
| E-1306 | 235 | 1285 |
| R-SJ21 | 116 | 860 |
| R-38 | 110 | 1045 |
| R-699 | 97 | 983 |
| 13 | 64 | 898 |
| R-33 | 60 | 448 |
| R-M1478 | 28 | 169 |
| R-M587 | 18 | 214 |
| R-M2561 | 15 | 221 |
| R-2151 | 14 | 300 |
| R-2145 | Trace | 205 |
| 19 | 0 | 400 |

*mg per 100 ml culture filtrate
(from reference 9)

TABLE 2: Slime Accumulation and Vegetative
Growth at 8 Days of Culture In
Monokaryons and Established
Dikaryons of Schizophyllum commune.

| Strain | Slime* | Mycelium* |
|---|---|---|
| 19 | 0 | 180 |
| R-693 | 267 | 598 |
| 19 x R-693 | 89 | 512 |
| 19 | 0 | 285 |
| 13 | 60 | 715 |
| 19 x 13 | 10 | 245 |
| E-714 | 222 | 987 |
| R-M1478 | 0 | 115 |
| E-714 x R-M1478 | 75 | 538 |
| 1 | 835 | 727 |
| R-MW46 | 745 | 423 |
| 1 x R-MW46 | 161 | 455 |
| 35 hf | 176 | 404 |
| 20 | 414 | 417 |
| 35 hf x 20 | 104 | 665 |
| R-699 | 36 | 440 |
| R-845 | 453 | 821 |
| R-699 x R845 | 58 | 586 |

*mg per 100 ml culture filtrate
(from reference 9)

Fig. 20. Effects of culture age on mycelial growth and slime accumulation in monokaryons vs. resultant dikaryon of *S. commune*. Left, strain 1 A51B41; middle, strain R-MW46 A43B43; right, resultant dikaryon. Solid line mycelium; dotted line, slime. (From ref. 9).

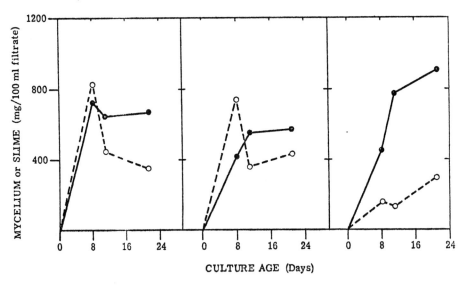

# REFERENCES

1.  Niederpruem, D. J. and R. J. Jersild (1972). Cellular aspects of morphogenesis in the mushroom *Schizophyllum commune*. CRC Critical Reviews in Microbiology 1:545-576.

2.  Sundberg, W. J. (1977). Hymenial cytodifferentiation in Basidiomycetes. In J. E. Smith and D. R. Berry, ed., The Filamentous Fungi III Fungal development. P.298-313. Edward Arnold (Publishers) Ltd., London.

3.  Niederpruem, D. J., R. A. Jersild and P. L. Lane (1971). Direct microscopic studies of clamp connection formation in growing hyphae of *Schizophyllum commune*. I. The dikaryon. Arch. Mikrobiol. 78:268-280.

4.  Mayfield, J. E. (1974). Septal involvement in nuclear migration in *Schizophyllum commune*. Arch. Microbiol. 95:115-124.

5.  Matthews, T. R. and D. J. Niederpruem (1972). Differentiation in *Coprinus lagopus*. I. Control of fruiting and cytology of initial events. Arch. Mikrobiol. 87:257-268.

6.  Matthews, T. R. and D. J. Niederpruem (1973). Differentiation in *Coprinus lagopus*. II. Histology and ultrastructural aspects of developing primordia. Arch. Mikrobiol. 88:169-180.

7.  Niederpruem, D. J. and S. Hunt (1967). Polyols in *Schizophyllum commune*. Amer. J. Bot. 54:241-245.

8.  Cotter, D. A. and D. J. Niederpruem (1971). Control of arabitol formation in *Schizophyllum commune* development. Arch. Mikrobiol. 78:128-138.

9.  Niederpruem, D. J., C. Marshall and J. L. Speth (1977). Control of extracellular slime accumulation in monokaryons and

resultant dikaryons of *Schizophyllum commune*. Sabouravidia 15:283-295.

10. Wessels, J. G. H. (1965). Morphogenesis and biochemical processes in *Schizophyllum commune*. Fr. Wentia 13:1-113.

11. Schwalb, M. N. (1977). Cell wall metabolism during fruiting of the Basidiomycete *Schizophyllum commune*. Arch. Microbiol. 114:9-12.

12. Sietsma, J. H., D. Rast and J. G. H. Wessels (1977). The effect of carbon dioxide on fruiting and on the degradation of a cell-wall glucan in *Schizophyllum commune*. J. Gen. Microbiol. 102:385-389.

# REGULATION OF FRUITING

Marvin N. Schwalb*

## INTRODUCTION:

The ability to provide biochemical and genetic analysis has made the fungi of fundamental importance in the study of the eukayotic cell. Fungal species have proven to be of particular utility in studies of differentiation and development. However, the utilization of higher fungi as model systems for studies of temporal development have not been fully explored.

In 1958, two reports appeared which demonstrated the potential of examining fruit body morphogenesis as a model of the genetic control of development. Esser and Staub (1) showed that a series of mutants could be identified which interfered with the normal development of *Sordaria,* and Raper and Krongelb (2) reported on an extensive analysis of fruiting in the basidiomycete *Schizophyllum commune.* Among other things, this latter work demonstrated that there was a complex genetic system controlling the formation of the relatively simple basidiocarp of *S. commune* and just as important, that this genetic system was amenable to analysis by standard techniques. In 1965, Wessels (3) examined a variety of parameters of fruiting in *S. commune* which demonstrated that fruiting was also amenable to biochemical analysis.

The development of fruit bodies of *S. commune* is divided into several morphologically distinct stages (Figure 1). In essence there are actually two developmental programs superimposed to make a complete fruit body. The five macroscopic stages as defined by Leonard and Dick (4) represent only three periods of differentiation, Stages I, II and III. Stages IV and V are phases of growth in the differentiated state in which there is a repeated, normally

* Department of Microbiology, College of Medicine and Dentistry of New Jersey, New Jersey Medical School, Newark, New Jersey 07103. The previously unpublished portions of this work were supported by Grant A109779 from the National Institute of Allergy and Infectious Diseases (U.S.).

asynchronous, sequence of cell differentiation leading to
basidiospore formation and ejection.

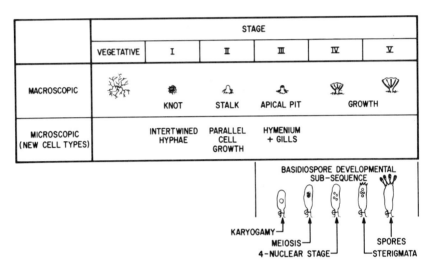

Fig. 1  Morphogenes is of fruit bodies in *Schizophyllum commune.*

Our studies of fruiting in *Schizophyllum* began, as all good
experiments should, by accident. While observing the mating
behavior of the recessive morphological mutant *thin,* we found that
in matings with normal strains, the *thin* mutant side of the mating
formed a relatively synchronous population of fruit bodies almost
immediately after dikaryotization (5). The technique was improved
by better control of $CO_2$ levels and temperature (6) so that we now
can obtain sufficient quantities of fruiting material at any stage of
development. This material is substantially free of any vegetative
dikaryon etc. (Fig. 2).

In species without a button stage such as *S. commune,* the
appearance of each stage is more readily visible and normally there is
no stipe growth. As we shall see later, any stipe growth which may
occur must precede the formation of the gilled stages in this species.
We believe that this synchronous technique has provided a new
experimental approach to the study of development of fruit bodies
in *S. commune.*

Fig. 2   Mature fruit bodies produced by the synchronous technique. 2X.

There are three principal environmental determinants of fruit body morphogenesis: light, $CO_2$ and gravity. The first two are of particular interest because they frequently control the ability of the fungus to undergo morphogenesis at all. Light may either inhibit or promote morphogenesis (7, 8) whereas high levels of $CO_2$ are frequently inhibitory (9, 10). The morphogenesis of fruit bodies of *S. commune* is an excellent example of the diversity and complexity of environmental responses (Figure 3).

| | | STAGE | | | | | |
|---|---|---|---|---|---|---|---|
| | | VEGETATIVE | I | II | III | IV | V |
| EFFECTOR | CO₂ | NO EFFECT | INHIBITION ————————————➤ | | | NO EFFECT | |
| | LIGHT | NO EFFECT | | REQUIRED ————————➤ | | NO EFFECT | |
| | | | | POSITIVE TROPISM | | | |
| | GRAVITY | NO EFFECT ————————————➤ | | | POSITIVE TROPISM | ASYMMETRICAL GROWTH | NEGATIVE TROPISM |

Fig. 3   Summary of the enviromental control of fruiting in *S. commune.*

Niederpruem et al. (11) investigated the nutritional aspects of fruiting and detemined that a wide variety of carbon and nitrogen sources will support fruiting. High levels of glucose (4% or more) suppress fruiting in *S. commune* (3, 5) as well as in *Coprinus* (12).

## BIOCHEMISTRY OF FRUIT BODY DEVELOPMENT:

In order to examine the regulatory processes which control fruiting it is necessary to detemine some of the biochemical changes associated with morphogenesis. These changes then serve as markers of the developmental process independent of the morphology.

The cell wall composition of *S. commune* has been examined in several laboratories (13,14,15). The major polymers are three glucans identified by their relative solubilities (WSG, water soluble glucan, mucilage, slime, extracellular polysaccharide; S glucan, alakali soluble glucan: R glucan, alakali insoluble glucan). Chitin is also present. Cell wall changes which occur during fruiting were first studied by Wessels (3). We recently reexamined this issue using the synchronous fruiting technique (16) which allows separation of fruit body from vegetative mycelium. Table 1 shows the ratios of the water-insoluble cell wall fractions in three wild-type strains. The S/R glucan ratio increases about 3 to 4 fold during fruiting. A fundamentally different pattern is found in the bug's ear (*bse*) mutation. This dominant fruiting abnormality whose morphology will be considered later, shows a *decrease* in the S/R glucan ratio during fruiting and the decrease is due to a relative increase in the R glucan-chitin fraction. The *bse* strains used are six generations coisogenic with strain 822-4. As a comparison of the published data shows, S/R ratios are dependent upon strain variation, culture conditions and the exact nature of the material analyzed (3,16,17). Basically, as Wessels has demonstrated (3), the same major polymers found in vegetative growth are found in fruit bodies, although a recent report (17) indicates that there is a change in the resistance of R-glucan to enzymatic hydrolysis during morphogenesis.

The quantitatively most important change in the cell wall during fruiting is the almost total loss of water soluble glucan. This polymer which may be 25-50% of the total dry weight in vegetative dikaryons is reduced to a trace quantity during fruit body morphogenesis. It should be noted, however, that WSG is not required for fruiting (3). We wished to determine where the WSG synthesis decreases during fruiting. Therefore, we pulsed 14C glucose during fruiting and determined the levels of uptake into cell wall

TABLE I.   Quantities of the water-insoluble cell wall polysaccharides
in strains of *Schizophyllum commune*.

| Phenotype | Stage[a] | R glucan chitin | S glucan | Ratio S/R |
|---|---|---|---|---|
| | | mg/mg protein | | |
| Wild-type[b] | VD | 4.6 | 0.6 | 0.13 |
| (1) | FBV | 3.0 | 1.8 | 0.59 |
| | VD | 2.5 | 0.6 | 0.24 |
| (2) | FBV | 2.6 | 2.2 | 0.85 |
| | VD | 6.5 | 1.0 | 0.15 |
| (3) | FBV | 1.6 | 0.7 | 0.43 |
| *bse* | VD | 3.0 | 2.5 | 0.83 |
| (1) | FBV | 4.1 | 2.3 | 0.56 |
| | VD | 3.4 | 2.5 | 0.71 |
| (2) | FBV | 4.8 | 2.5 | 0.52 |
| | VD | 2.5 | 1.9 | 0.76 |
| (3) | FBV | 4.9 | 2.3 | 0.47 |

[a]VD = vegetative dikaryon FBV = mature fruit body

[b]wild-type strain (1) is 6 generations coisogenic with the three *bse* strains.

Data adapted from (16)

polymers. Table 2 shows that the WSG synthesis decreases between stages II and III. R and S glucan synthesis continue at maximal levels throughout fruiting.

TABLE 2. Uptake of $^{14}$C glucose into cell wall fractions during development of fruit bodies of *Schizophyllum commune*

| Period of exposure | Specific activity (cpm/mg) | | |
|---|---|---|---|
| | R glucan | S glucan | WSG |
| Stage I - V | 2700 | 2300 | 1800 |
| Stage II - V | 2150 | 2120 | 1280 |
| Stage III - V | 2325 | 2320 | 360 |

From reference 16.

The role of WSG in *S. Commune* is not well understood. Wang and Miles (18) provided evidence that WSG may be a reserve carbon source. However, in starvation experiments we find no rise in the enzymes which degrade WSG until after autolysis of protein has begun. We also have evidence that WSG is not a capsule material but is located firmly in the cell wall. When vegetative mycelium grown on cellophane membrane supports are broken by gentle cutting, 50% of the cell protein is released without any significant quantity of WSG. Furthermore, when mycelia grown in this fashion are vigorously shaken in buffer, the readily soluble WSG is not removed from the wall. These results suggest that the "extracellular" nature of WSG noted previously (14) may be an artifact of the growth in liquid media.

We also wished to characterize some of the enzyme changes associated with fruit body morphogenesis. Besides their intrinsic interest, such changes serve as markers of the time course of development in comparative studies with mutants. Figure 4 shows the changes in activity of our three most important "marker" enzymes, phosphoglucomutase (19), glucoamylase (20) and inactivating protease (21). The latter two enzymes normally only appear in the fruit body, the glucoamylase appearing immediately after fruiting begins and the protease is first detected at stage II.

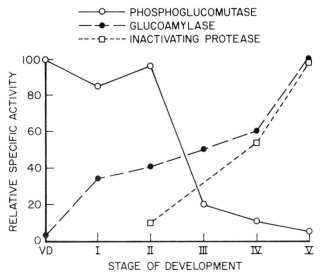

Fig. 4 The pattern of activity of three enzymes during the development of fruit bodies.

The activity of the mutase and glucoamylase represent an interesting example of regulation of metabolic activity during development. Figure 5 shows the pathway of glucose utilization during development in *S. commune.* At stage II or later the only enzymes in this pathway which show any substantial changes in specific activity are phosphoglucomutase and glucoamylase. Since we cannot detect any glycogen phosphorylase in this organism, all glycogen degradation produces glucose which must then be phosphorylated. In this way all G-I-P is synthesized by phosphoglucomutase ensuring that the mutase alone controls the flow of substrate into the pathway of glucan synthesis. The changing relationship between the three enzymes of glucose-6-phosphate utilization suggests that during the vegetative stage, the level of substrate (glucose-6-phosphate) is the limiting factor and in fruiting the level of enzyme (phosphoglucomutase) becomes the limiting factor in the supply of substrate for glucan synthesis.

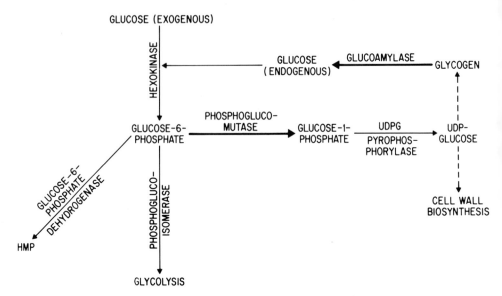

Fig. 5 Pathway of glucose-1-phosphate synthesis during fruting. Thick arrows
indicate those enzymes whose activity shows substantial change during
development.     (see Fig. 4)

The inactivating proteases are unique enzymes. They were
originally observed as part of a study of the cause of the loss in
phosphoglucomutase activity. We found that during fruiting an
enzymatic activity occurred which modified the mutase so that is
showed an increase in cold lability (22). Further examination
demonstrated that this modifying activity was due to several
chromatographically separable proteases which also have the capacity
to inactivate a number of *S. commune* enzymes (21). In this respect
the proteases are similar to the "B" - type protease of yeast (23),
except that the *S. commune* enzymes are specific for the
developmental phase.

The proteases were purified by CM-cellulose chromatography.
Table 3 shows the activity of the protease(s) against various test
enzymes. The proteases are sensitive to a serine protease inhibitor
and are active in the neutral to weakly alkaline pH range. Most
interesting of all is the evidence that the proteases only provide
limited breakdown of substrates. The evidence is as follows:

(a) The activity against *S. commune* phosphoglucomutase is apparently very slight so that only cold lability of the mutase is affected.

(b) The purified proteases show very little activity in releasing TCA soluble products from denatured hemoglobin at pH3 or casein at pH 7.

(c) Chromatographic analysis on Sephadex gels shows that the breakdown products of the proteases from high molecular weight substrates such as hemoglobin and rabbit muscle PGM are still of relatively high molecular weight.

The common role ascribed to this type of protease in an initial discriminating attack as part of protein turnover, although similar enzymes are suspected of causing zymogen activation as part of morphogenesis in yeast (24). Since we have been unable to detect any inactivating protease in vegetative stages, a problem arises. The substrate enzymes are undoubtedly turned over in vegetative cells as well as fruiting material. Either the proteases are a new mechanism of turnover specific for development or they function in some other area of metabolism. Unless there is some *in vivo* specificity which we do not as yet understand, the random spectrum of *in vitro* specificity suggests no pattern and I suspect that these proteases play an as yet undetermined role in development. Although it is tempting to equate the protease with the loss of mutase, we do not have any direct proof of this relationship.

Other enzymes which are known to be regulated during fruiting include glucose-6-phosphate dehydrogenase (19), caseinase, trehalase (25) and polyphenol oxidase.

The polyphenol oxidase system which has been studied by Leonard and his coworkers (26,27) shows a complex pattern of intra-and extracellular enzymes. Their results suggest that an inhibitor is produced which is responsible for the decline in activity found in mature fruit bodies.

CYCLIC AMP AND MORPHOGENESIS:

Two other enzymes which may be regulated during the fruiting process are of particular note. These are adenyl cyclase and cyclic

TABLE 3.  Inactivating activity of *S. commune* protease
against various test enzymes

| Substrate enzyme | Relative activity | |
|---|---|---|
| | Crude protease | Peak 3 protease |
| S. commune | | |
| UDP-glucose pyrophosphorylase | 220 | 90 |
| Isocitric dehydrogenase | 170 | 120 |
| 6-Phosphogluconate dehydrogenase | 100 | 100 |
| Phosphoglucomutase | 85 | 90 |
| Malic dehydrogenase | 21 | 25 |
| Phosphoglucoisomerase | 19 | 53 |
| Glucose-6-phosphate dehydrogenase | | 68 |
| Trehalase | 0 | |
| Glucoamylase | 0 | |
| Rabbit muscle | | |
| Phosphoglucomutase | 200 | 300 |
| Pig heart | | |
| Isocitric dehydrogenase | 45 | 49 |
| Malic dehydrogenase | 7 | 0 |
| Yeast | | |
| 6-Phosphogluconate dehydrogenase | 40 | 86 |
| Phosphoglucoisomerase | 0 | |

The activities are expressed as relative to the amount of activity toward *Schizophyllum commune* 6-phosphogluconate dehydrogenase (100%). Each protease sample was considered independently. The initial specific activities are 5.4 for the crude extract and 139 for the Peak 3 proteases.  (From ref. 21)

AMP phosphodiestrase in *Coprinus macrorhizus*. This observation was made as part of the very important work of Uno and Ishikawa on the role of cyclic AMP in fruiting of *C. macrorhizus*.

Although review of their work has recently appeared (28), no discussion of the regulation of fruiting would be complete without some mention of their findings. The level of cyclic AMP in *C. macrorhizus* is directly related to fruit body formation. Cyclic AMP acts as a fruit inducing substance and only accumulates during the fruiting process. In addition to the adenyl cyclase and phosphodiesterase, three classes of protein kinases also appear during fruiting. The cyclic AMP can remove the catabolite repression of fruiting and enzyme induction by high levels of glucose. In summary, the ability to increase the level of cyclic AMP appears to be a necessary part of fruit body induction in *Coprinus*.

Since a fruit inducing substance has been observed in other species (29,30) including *S. commune* (4), we decided to see if cyclic AMP plays a similar role in this species. After examining a wide variety of concentrations we found no evidence that cyclic AMP, the dibutyl derivative, or cyclic GMP would act as a fruiting inducer. However, we discovered that cyclic AMP could profoundly affect the morphology of developing fruit bodies (31). At $10^{-3}$M concentration in regular fruiting medium, cyclic AMP causes the formation of small, gill-less fruit bodies (Figure 6) in the wild type strains. Since this phenotype is precisely that of the *bse* mutant, we also examined the effects of exogenous cyclic AMP on *bse* strains (32). In strains heterozygous for *bse,* the cyclic AMP causes the complete loss of all of the cellular and macroscopic characteristics of stage II. Instead, smooth, spore-producing hymenial areas are formed on relatively undifferentiated masses of tightly knit hyphae. Further studies have shown that just as the cyclic AMP causes the wild-type strains to phenocopy heterozygous *bse,* it also causes strains heterozygous for *bse* to phenocopy strains homozygous for *bse*. Therefore, exogenous cyclic AMP can precisely mimic the effects of the *bse* mutation at differing gene dosage.

The inference that the *bse* gene is involved in cyclic AMP metabolism was obvious and we investigated this situation using the radioisotipe dilution assay with cyclic AMP specific binding protein.

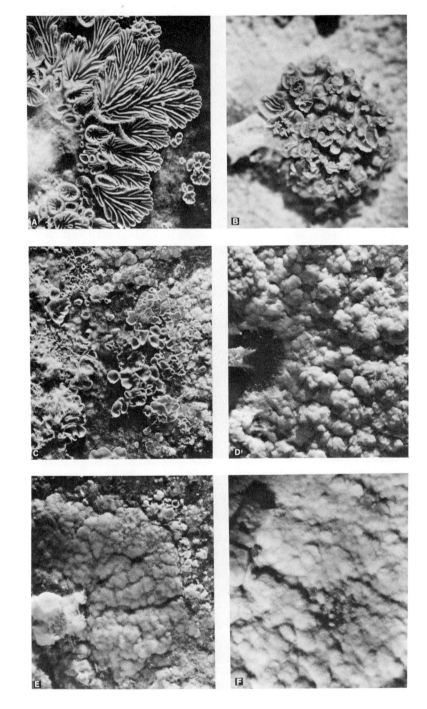

Fig. 6   The morphological effects of cyclic AMP on wild-type and *bse* mutant
         fruit bodies
         A. Wild-type, no treatment.
         B. Wild-type grown with 1mM cyclic AMP.
         C. *bse* x wild-type, no treatment
         D. *bse* x wild-type grown with 1mM cyclic AMP
         E. *bse* x *bse* no treatment
         F. *bse* x *bse* grown with 1mM cyclic AMP. (From ref. 32)

Our preliminary results shown in Figure 7 demonstrate that cyclic
AMP levels are elevated in strains carrying the *bse* mutant. The sum
of these data also suggest that the elevation of cyclic AMP levels
itself is not adequate to explain the entire *bse* effect.

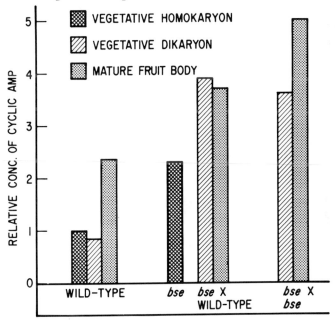

Fig. 7   Relative concentration of cyclic AMP in *S. commune.*
         One unit equals approximately 10 p mole/mg protein.

In *S. commune*, cyclic AMP can control morphogenesis of fruit
bodies and the *bse* mutant appears to cause constitutive, positively
regulated synthesis of the cyclic AMP.

These results along with those observed in *Coprinus* and other
fungal species clearly demonstrate that cyclic AMP may be a basic
regulator of morphogenesis in the fungi.

A point which is still to be answered however, is the role of cyclic AMP as an initiator of the fruiting response in *S. commune.* It is of obvious utility to be able to identify the fruit inducing substance. As I noted our original results with the application of exogenous cyclic AMP were negative. However, during the course of our program to obtain coisogenic strains for comparative studies, we found a rather surprising result, although in retrospect it is not so surprising after all. By the time we got to the fourth back-crossed generation of *bse* we found that *bse* monokaryons fruited regularly (Figure 8). A preliminary direct analysis of this situation showed that

Fig. 8    Monokaryotic fruiting in a strain carrying the *bse* mutation. 2X.

100% of the *bse* isolates in a cross with the wild-type strain with which it is coisogenic were monokayotic fruiters. None of the wild-type isolates showed monokayotic fruits. Because of the variability inherent in monokayotic fruiting in wild-type strains of *S. commune* the negative aspect of the result, no fruiting in wild-type, is not absolute. However, it is more of a coincidence than this observer can bear to assume that the results are due to two different closely linked cistrons. Rather the pattern of high endogenous levels of cyclic AMP associated with *bse,* when considered with the results with *Coprinus,* clearly indicate that cyclic AMP is a fruiting initiator in *S. commune* as well. These results present two other problems.

First, why didn't the original *bse* strain show homokayotic fruiting? The answer to this lies in the process of making the *bse* strain coisogenic. As we processed each generation we selected not only the *bse* phenotype itself but those examples which showed the best fruiting potential. Therefore, we were also indirectly eliminating those recessive mutants deleterious to fruiting which regularly accumulate in aged dikaryons(33). As proof of this we noted that at the first generation, *bse* x *bse* mating produced no fruit bodies at all, but by the fourth generation, the distinctive *bse* x *bse* fruit body regularly appeared. The second question is more difficult to explain. If the presence of an exogenous inducer is clearly identifiable in *S. commune* (4), then why doesn't cyclic AMP work. I can only suggest that there may be problems of uptake, binding etc. and that another related compound is serving to mimic the effect. This issue also raises a semantic problem. It is important that the term "fruit inducing substance" (fis) be clearly limited to exogenously produced effects. This type of response may be fundamentally different from a fruiting initiator which I am using to describe an endogenous substance. The evidence from both *Schizophyllum* and *Coprinus* suggests that these substances may not always be the same.

GENETIC CONTROL OF FRUITING:

We have already touched upon some aspects of the genetic control of fruiting. One of the more vexing problems in the control of fruit body initiation is the question of the role of the incompatibility factors. Even though mutants have been isolated in several species which permit monokayotic fruiting, clearly the establishment of the dikaryon is a trigger to fruiting initiation. Recently, Stahl and Esser (34) examined this problem in *Polyporus ciliatus*. They provided evidence that the B factor plays a role in inhibiting monokaryotic fruiting. Basically there are two possibilities which are not necessarily mutually exclusive. Either the mating type factors inhibit monokaryotic fruiting (negative control) or the mating type factor interaction in dikaryosis promotes fruiting (positive control). If the mating type factors inhibit monokaryotic fruiting one would expect that at least some of the monokaryotic fruiting mutants would be related to the interference with A or B factor function but this has not been demonstrated. Rather the type II modifers of A and B factor function in *S. commune* inhibit fruiting in dikaryons(35). Furthermore, there are two monokaryotic fruiting genes in which something is known of their biochemistry. The *fis*[c] gene in *Coprinus* and the *bse* gene in *Schizophyllum* both involve the ability to elicite a cyclic AMP activity in monokaryons where it is normally not found. Also in *Coprinus* the ability to be induced by cyclic AMP is apparently a mutant condition in that only one of the 16 isolates studied was *fis*[c]. The sum of these scattered observations suggest that the establishment of the dikaryon *promotes* fruiting and that monokaryotic fruiting arises by the acquisition of some activity normally activiated by the mating factor interaction. The evidence in *S. commune* on the increased fruiting potential in B mut strains would implicate that factor in this species (36). It should also be noted that although monokaryotic fruiting may appear morphologically identical to dikaryotic fruiting, in *C. macrorhizus* the *fis*[c] mutant has a different response to light than dikaryotic fruits. In addition, it is clear that genes at several different loci in *S. commune* can be involved in monokaryotic fruiting as part of their phenotypic expression.

Raper and Krongelb analyzed fruiting competence in a large sample of dikaryons. Although at that time, the techniques for optimizing the environmental conditions for fruiting were not known, the basic conclusions that the ability to fruit is a genetically complex characteristic and that some strains are genetically poor fruiters appear to be still valid. Certainly, even under optimum conditions it is not uncommon to find poor fruiting strains of *S. commune*. As noted, fruiting competence is probably related to the accumulation of a diverse group of deliterious mutants. There are, however, more specific genetic units which prohibit fruiting such as the *thin* mutants of *S. commune* (5) and the "Knotless" mutant of *Coprinus* (37).

A wide variety of mutants such as *bse* are known to interfere with normal development after initiation. The single gene *bse* mutant has a pleotropic phenotype (Table 4). This situation is not uncommon in morphological mutants and probably reflects the interlocking metabolic control of morphology. Besides *bse* the *cor* (2) characters also shows semi-dominance. The pattern suggest that each haploid genome produces only half of the total regulatory proteins required for normal control of fruiting. This type of control is similar to the *nir* locus of *Aspergillus nidulans* (38).

TABLE 4. Phenotypic changes associated with the *bse* mutant

1.   Semi-dominance

2.   Abnormal morphogenesis: no gills in single dose and no stage II or gills in double dose

3.   Abnormal cell wall composition

4.   Abnormal cyclic AMP metabolism

5.   Abnormal response to exogenous cyclic AMP

6.   Monokaryotic fruiter

7.   Constituative synthesis of glucoamylase

8.   Decreased sensitivity to inhibition by 5% $CO_2$

SPORULATION:

As I noted in the beginning, fruiting is really two developmental processes, the development of the macroscopic structure and the process of sporulation. I will consider some of the evidence for this separation later. For now, suffice to note that sporulation differs from the activities we have noted so far in one important aspect. It is an act of differentiation confined to a single cell rather than the coordinate differentiation of multicellular development. And as I will try to demonstrate in a moment, this is true even in those species where sporulation is a single synchronous act. $CO_2$ has no effect on sporulation in *S. commune* although light is required. Mature fruit bodies placed in the dark, stop sporulating (39), and the basidia become inhibited at some point in the one-nuclear stage. Similar light related effects have been observed in several species of *Coprinus* (40,41). Figure 9 shows the recovery from dark inhibition. The

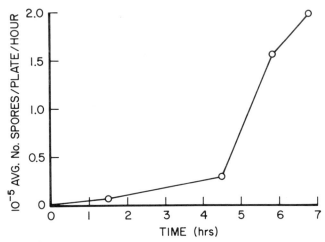

Fig. 9   Recovery from the dark-inhibition of sporulation.
Cultures are placed in light at zero time.

recovery time of 5-6 hours indicates the period between early meiosis and spore ejection. Furthermore, the dark-light change produces a synchronous burst of sporulation. Therefore, a simple environmental control can coordinate the activity of each basidium without any apparent multicellular, internal coordination.

We have also been able to isolate several temperature sensitive sporulation-less mutants (42). Two of these are of particular note. The mutant *spo* is wild-type at 20°. At 30° the fruit body is wild-type but there is no sporulation and many basidia show meiotic figures. Temperature switch experiments show that the recovery from the restrictive conditions with *spo* requires between 2-4 hours which indicate that meiosis is not completed before this period of time in the sporulation cycle.

If I might be permitted a totally unscientific statement, the other sporulation-less mutant, *stg,* is my all time favorite mutant. At the restrictive temperature, *stg* forms abnormal sterigmata (Figure

10). These sterigmata lack the normal apical determinant and appear more like vegetative hyphae. Furthermore, there are frequently fewer than four sterigmata formed. Those persons concerned with the evolution of the Basidiomycetes will note the similarity to certain

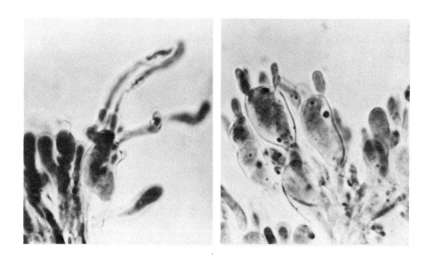

Fig. 10 The basidia of the *stg* mutant of *S. commune* at the restrictive temperature, 30° (41).

members of the heterobasidiomycetes. Both *spo* and *stg* are single gene recessive characters. *Stg* appears to have lost control over the unique limitation of apical growth which characterizes sterigmata development. What is perhaps most interesting however, is the nuclear condition of the *stg* phenotype. Since the development of a functional technique for straining nuclei and meiotic figures in *S. commune* by Radu, Steinlanf and Koltin (43) we are able to regularly observe nuclear behavior in basidia. However, upon examining several hundred cells exhibiting the *stg* phenotype, we have been unable to find any nuclear or meiotic figures at all in these cells. The clear suggestion here is that the nuclei are inhibited at one of the stages of meiosis which do not show any meiotic figures even with staining. This is not consistent with the repeatedly observed sequence of sterigmata development occuring after the four nuclear stages (44,45). Apparently the signal for the initiation of sterigmata is started prior to their appearance and during an earlier part of the differentiative sequence and even though meiosis is never completed, the information to initiate the sterigmata continues to be processed. A similar situation has been found in the mitotic cycle in yeast (46).

REGULATION OF FRUITING:

Our studies of fruiting in *S. commune* have allowed for the development of certain concepts about the fruiting process in this organism. The applicability of these results to other species depends of course, on the degree to which the results are specific to the needs of a particular fungus. Obviously as one progresses to more molecular issues the results become more universally applicable.

In *S. commune,* by optimizing the environmental conditions one can obtain excellent fruiting without starvation or mobilization of reserves. This situation may be somewhat artificial in that when there is a more balanced mix of vegetative and fruiting material Wessels has shown that pilei will start to grow after all the glucose (and earlier all the nitrogen) have disappeared from the medium. In this situation, the fruit bodies can continue to grow by drawing upon a reserve substance such as R-glucan and/or WSG. This is where R-glucanase activity would be of importance (47). Also, in the absence of a nitrogen source, pileus growth is inhibited by high concentrations of glucose (3). The sum of these observations suggest that the nutritional control of pileus growth and expansion is a complex phenomenon which is not fully understood. Therefore, while a sustained low concentration of metabolizable carbohydrate is necessary under certain conditions, this is not always a requirement.

There is an inverse relationship between vegetative growth and fruiting. Either fruiting inhibits vegetative growth or the *promotion* of vegetative growth inhibits fruiting. Carbon dioxide acts as one mediator of this relationship in *S. commune.* It is conceivable that $CO_2$ inhibits fruiting by promoting vegetative growth, rather than fruit bodies having a $CO_2$ sensitive metabolism. In relationship to this, starvation requirements in some species could be interpreted as a mechanism for inhibiting vegetative growth to allow fruiting to occur.

These observations raise a fundamental issue concerning the regulation of development in organisms, such as many fungi, where the development has two basic interrelated parts: growth (i.e. cell and nuclear division and/or cell extension) and differentiation. Differentiation only occurs during growth, but growth may occur within a differentiated state without new differentiation.

In *S. commune* there is normally relatively little growth during the phases of differentiation when the environmental conditions are optional. That is, when light and $CO_2$ levels are correct, the sequence from Stage I to III occurs with relatively little growth. In strains carrying the *med* character (and in normal strains as well, although it is not as easily observed) this sequence can be interrupted by suboptional environmental conditions to produce extensive growth at Stage II, i.e. a long stalk (48). When growth occurs at Stage II, Stage III does not appear but this in itself is of no significance since the growth at Stage II is apical. However, this growth at Stage II can be interrupted at any time and Stage III made to appear by optimizing the environment (Figure 11). Furthermore, the clearly growth stages of macroscopic development, Stage IV and V are $CO_2$ and light insensitive. *Therefore, it becomes possible to separate and define the phases of differentiated growth at those which are sensitive to certain environmental stimuli* and the period which are strictly growth without differentiation as those which are insensitive to these stimuli. This hypothesis can be a usefut tool in understanding the interrelationship of the growth and differentiation phases of development, at least in this species.

With this as a background we can begin to consider the molecular aspects of regulation. We already know that growth in the differentiated state is not limited but controlled by external effectors. We have made stalks of *S. commune* that were 5 cm. long in *med* strains. Then what is the relationship between growth *per se* and differentiated growth? We can approach this question by examining some of the changes in biochemistry which appear to take place at the transistion between Stage II and III. The level of PGM in the *med* mutant is shown in Figure 12. The loss in activity occurs not as a result of differentiation to Stage III but during Stage II. In addition glucomylase and protease continue to accumulate. That is,

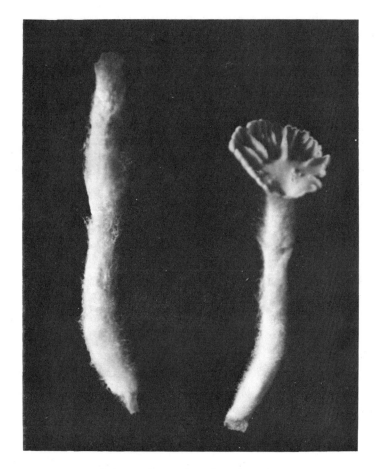

Fig. 11    Examples of the *med* mutation phenotype.
           Left: Long stalk; Right: Completion of morphogenesis upon the
           long stalk.

the event which is normally associated with a sequence of
differentiation is in fact a single signal which continues even during
growth without differentiation. Furthermore, if we inhibit
development at Stage II with $CO_2$ a situation which also inhibits
growth, no change in the activity of PGM or glucoamylase is
observed. Therefore, the response is due to the continuation of
growth, not aging. When differentiation is reiniatiated, there is an
increase in PGM in the Stage III material, assumedly new synthesis,
and a repetition of the sequence. The inference is that the molecular
controls follow a time course which is independent of morphology

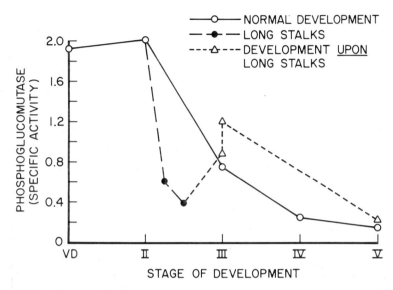

Fig. 12    Pattern of phosphoglucomutase activity in strains with the *med* phenotype.

*per se.* That is, time is an independent variable.

In cyclic AMP treated *bse* strains, hymenium is produced on masses of hyphae having none of the morphological characteristics of Stage II. However, the level of PGM and all other identifiable biochemical changes occur in the cyclic AMP treated *bse*. Therefore, the signal for the loss of PGM is not dependent on having a Stage II at all. It appears as if morphology is an *indirect* consequence of the controlling temperal sequence.

When we originally began this work, we asked the question: What are the controls which coordinate events so that all of the metabolic changes required to produce a new morphology arrive at the same place at the required time? These data suggest that for some parts of the process, there are no interlocking controls. Signals are sent for several changes which are required at some later time. Very general molecular controls over the expression of the changes are timed internally and independently. They arrive at a common point in time without ever keeping track of each other. So, if you interfere with the normal sequence, the expression of the characteristics continue. Then when differentiation is allowed to proceed, if the level of product is not correct, the sequence may be recapitulated. Remember the situation with the *stg* mutant where the signal for the event appears to proceed the event in time by a substantial amount and the expression of the signal does not require the normally interveneing events to take place. The changes in enzyme activity in cyclic AMP treated *bse* appears to be a related phenomenon.

Figure 13 demonstrates our concept of the relationship of growth and differentiation in *S. commune*. The figure shows that it is possible to conseptually separate differentiative growth and growth in a differentiated state.

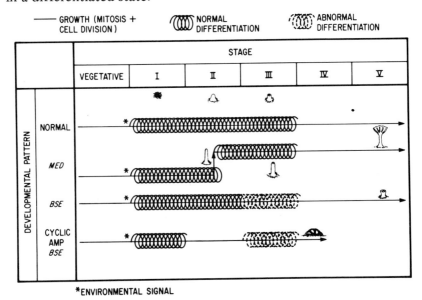

Fig. 13    Relationship of growth and differentiation during fruiting of *S. commune.*

In summary, the results of these studies on fruiting in the higher basidiomycetes have yielded several hypotheses:

1. There is a basic relationship between growth and differentiation. The relationship suggests that there is a fundamental difference between vegetative and differentiative growth. In the higher fungi this difference manifests itself by a change in the response to environmental effectors. Although we often say that "$CO_2$ inhibits fruiting" in fact it does not. It inhibits growth of fruiting cells. It is possible that in

differentiative growth, the cell cycle itself is fundamentally altered so that it is the cell cycle which is effected by environment and not differentiation *per se.* If this is true, it indicates a basic difference in the regulation of these aspects of morphogenesis, such as sporulation, which occur after the phase of differentiative growth. Sporulation, for example, only involves escentially two cell divisions and the coordination between cells is not required. Therefore, the act of final *differentiative growth* is the acquisition by sub-hymenial cells of the ability to form basidia; and the formation of basidia and spores is part of growth in the *differentiative state.*

2.  The genetic regulatory elements of development in the higher fungi appear to be similar to those which control vegetative metabolism. Dosage effects may be common and probably related to the essentially diploid nature of this part of the life cycle.

3.  The time sequence of events indicates that each morphological event is not required for the next nor for the expression of certain metabolic events. As a corallary to this, some metabolic (genetic?) events appear to proceed as a response to continued growth rather than differentiation. This again suggests that cell division in the differentiated state may itself be a regulator of differentiation. As long as differentiated growth continues, programmed changes in enzyme activity also continues.

4.  Cyclic AMP plays a fundamental role in the regulation of development in the higher fungi. The ability to initiate a change in cyclic AMP metabolism may be the missing ingredient in monokaryons for fruit development. In this connection it is important to note that cyclic AMP is believed to play a basic role in the regulation of the cell cycle (49).

Finally, beyond these practical matters I would like to leave you with this thought: "...the fruiting process is revealed as a far more complex phenomenon that has been previously recognized. These findings reported here would seem to indicate a new dimension for future inquiry on fruiting in the Hymenomycetes". This last statement is not mine but was made by John Raper almost twenty years ago (2).

# REFERENCES

1.  Esser, K. and J. Staub, 1958. Genetishe Untersuchungen an *Sordaria macrospora* Auersw., Kompensation und Induktion bei genbedingten Entwicklungsdetekten. Z. Vererebungsl. **89**:729.

2.  Raper, J.R. and G. Kongelb. 1958. Genetic and environmental aspects of fruiting in *Schizophyllum commune.* Mycologia **30**:707.

3.  Wessels, J. G. H. 1965. Morphogenesis and biochemical processes in *Schizophyllum commune* Fr. Wentia **13**:1.

4.  Leonard, T. J. and S. Dick. 1968. Chemical induction of haploid fruit bodies in *Schizophyllum commune.* Proc. Nat. Acad. Sci. (U.S.) **59**:745.

5.  Schwalb, M. N. and P. G. Miles. 1967. Morphogenesis of *Schizophyllum commune.* I. Morphology and mating behavior of the thin mutation. Amer. J. Bot. **54**:440.

6.  Schwalb, M. N. 1971. Commitment to fruiting in synchronously developing cultures of the basidiomycete *Schizophyllum commune.* Arch. Mikrobiol. **79**:102.

7.  Perkins, J. H. 1969. Morphogenesis in *Schizophyllum commune* I. Effect on white light. Plant Physiol. **44**:1706.

8.  Morimoto, N. and Y. Oda. 1973. Effects of light on fruit body formation in a basidiomycete, *Coprinus macrorhizus.* Plant and Cell Physiol. **14**:217.

9.  Niederpruem, D. J. 1963. Role of carbon dioxide in the control of fruiting of *Schizophyllum commune* J. Bacteriol. **85**:1300.

10. Tschierpe, H. J. and J. W. Sinden. 1964. Weitere

untersuchungen uber die Bedeutung von Kohlendioxyd fur die Fruktifikation des Kulturchampignons *Agaricus campestrus,* var. *bisporus* (L)Lge. Arch. Mikrobiol. **49**:405.

11. Niederpruem, D. J., H. Hobbs and L. Henry. 1964. Nutritional studies of development in *Schizophyllum commune.* J. Bacteriol. **88**:1721.

12. Tsusue, Y. M. 1969. Experimental control of fruiting body formation in *Coprinus macrorhizus.* Develop. Growth and Different. **11**:164.

13. Sietsma, J. A. and J. G. H. Wessels. 1977. Chemical analysis of the hyphal wall of *Schizophyllum commune.* Biochim. Biophys. Acta **496**:225.

14. Wessels, J. G. H., D. R. Kreger, R. Marchant, B. Regensburg and O. H. H. De Vries. 1972. Chemical and morphological characterization of the hyphal wall surface of the basidiomycete *Schizophyllum commune.* Biochim. Biophys. Acta **273**:346.

15. Siehr, D. J. 1976. Studies on the cell wall of *Schizophyllum commune.* Permethylation and enzymic hydrolysis. Can. J. Biochem. **54**:130.

16. Schwalb, M. N. 1977. Cell wall metabolism during fruiting of the basidiomycete *Schizophyllum commune.* Arch. Microbiol. **114**:9.

17. Siestma, J. H., D. Rast and J. G. H. Wessels. 1977. The effect of carbon dioxide on fruiting and the degradion of a cell-wall glucan in *Schizophyllum commune.* J. Gen. Microbiol. **102**:385.

18. Wang, C. S. and P. G. Miles. 1964. The physiological characterization of dikaryotic mycelia of *Schizophyllum commune.* Plant Physiol. **17**:573.

19. Schwalb, M. N. 1974. Changes in activity of enzymes metabolizing glucose-6-phosphate during development of the basidiomycete *Schizophyllum commune*. Devel. Biol. **40**:84.

20. Schwalb, M. N. 1971. Developmental regulation of amylase activity during fruiting of *Schizophyllum commune*. J. Bacteriol. **108**:1205.

21. Schwalb, M. N. 1977. Developmentally regulated proteases from the basidiomycete *Schizophyllum commune*. J. Biol. Chem. **252**:8435.

22. Schwalb, M. N. 1975. Developmental control of enzyme modification during fruiting of the basidiomycete *Schizophyllum commune*. Biochem. Biophys. Res. Comm. **67**:478.

23. Holzer, H., H. Betz and E. Ebner. 1975. Intracellular proteinases in microorganisms. Curr. Topics Cell. Reg. **9**:103.

24. Cabib, E., R. Ulane and B. Bowers. 1974. A molecular model for morphogenesis: The primary septum of yeast. Curr. Topics Cell Reg. **8**:1.

25. Bromberg, S. K. and M. N. Schwalb. 1978. Sporulation in *Schizophyllum commune:* Changes in enzyme activity. Mycologia. **In press.**

26. Leonard, T. J. 1971. Phenoloxidase activity and fruiting body formation in *Schizophyllum commune*. J. Bacteriol. **106**:162.

27. Phillips, L. E. and T. J. Leonard. 1976. Extracellular and intracellular polyphenoloxidase activity during growth and development in *Schizophyllum*. Mycologia **68**:268.

28. Ishikawa, T. and I. Uno. 1977. A mechanism of fruiting body formation in basidiomycetes. In: Growth and differentiation in

microorganisms. T. Ishikawa, Y. Maruyama and H. Matsumiya eds. pg. 283, University Park Press, Balt.

29. Urayama, T. 1957. Preliminary note on a stimulative effect of certain specific bacteria upon fruit body formation in *Psilocybe panaeoliformis* Murrill. Bot. Mag. (Tokyo) **70**:29.

30. Urayama, T. 1969. Stimulative effects of extracts from fruit bodies of *Agaricus bisporus* and some other hymenomycetes on primordia formation in *Marasmius* sp. Trans. Mycol. Soc. Japan **10**:73.

31. Schwalb, M. N. 1974. Effect of adenosine 3',5'-cyclic monophosphate on the morphogenesis of fruit bodies of *Schizophyllum commune*. Arch. Microbiol. **96**:17.

32. Schwalb, M. N. 1978. A developmental mutant affecting 3':5'-cyclic AMP metabolism in the basidiomycete *Schizophyllum commune*. FEMS Microbiol. Letters: **3**:107.

33. Miles, P. G. 1964. Dikaryons and mutations. Bot. Gaz. **125**:301.

34. Stahl, U. and K. Esser. 1977. Genetics of fruit body production in higher basidiomycetes. I. Monokaryotic fruitings and its correlation with dikaryotic fruiting in *Polyporus ciliatus*. Molec. Gen. Genet. **148**:183.

35. Raper, J. R. 1966. Genetics of sexuality in higher fungi. Ronald Press, N. Y.

36. Takamaru, T. and T. Kamada. 1972. Basidiocarp development in *Coprinus macrorhizus*. I. Induction of developmental variations. Bot. Mag. (Tokyo) **85**:51.

37. Cove, D. J. 1969. Evidence for a near limiting concentration of a regulator. Nature **224**:272.

38.  Bromberg, S. K. and M. N. Schwalb. 1976. Studies on basidiospore development in *Schizophyllum commune*. J. Gen. Microbiol. **96**:409.

39.  Lu, B. C. 1970. Dark dependence of meiosis at elevated temperatures in the basidiomycete *Coprinus lagopus*. J. Bacteriol. **13**:833.

40.  Lu, B. C. 1974. Meiosis in *Coprinus*. V. The role of light on basidiocarp initiation, mitosis and hymenium differentiation. Canad. J. Botany **52**:229.

41.  Bromberg, S. K. and M. N. Schwalb. 1977. Isolation and characterization of temperature sensitive sporulationless mutants of the basidiomycete *Schizophyllum commune*. Can. J. Genet. Cytol. **19**:477.

42.  Radu, M., R. Steinlauf and Y. Koltin. 1974. Meiosis in *Schizophyllum commune*. Arch. Microbiol. **98**:301.

43.  Erlich, H. G. and McDonough, E. S. 1949. The nuclear history in the basidia and basidiospores of *Schizophyllum commune* Fries. Am. J. Bot. **36**:360.

44.  Wells, K. 1965. Ultrastructural features of developing and mature basidia and basidiospores of *Schizophyllum commune* Fr. Mycologia **57**:236.

45.  Hartwell, L. H. 1974. *Saccharamyces cerevisiae* cell cycle. Bact. Rev. **38**:164.

46.  Wessels, J. G. H. 1966. Control of cell-wall glucan degradation during development in *Schizophyllum commune*. Ant. von Leeuwen. **32**:341.

47.  Schwalb, M. N. and A. Shanler. 1974. Phototropic and geotropic responses during the development of normal and

mutant fruit bodies of the basidiomycete *Schizophyllum commune.* J. Gen. Microbiol. 82:209.

48. Abell, C. W. and T. M. Monahan. 1973. The role of 3',5'-cyclic monophosphate in the regulation of mammalian cell division. J. Cell Biol. **59**:549.

# Index